FIX ON THE RISING SUN

The Clipper Hi-jack of 1938
—and the Ultimate M-Z

新潮選書

FIX ON THE ☀ RISING SUN

The Clipper Hi-jacking of 1938
—and the Ultimate M.I.A.'s

by

Charles N. Hill

Congress shall make no law respecting an establishment of religion, or prohibiting the free exercise thereof; or abridging the freedom of speech, or of the press, or the right of the people peaceably to assemble, and to petition the Government for a redress of grievances.

—Amendment I to the U.S. Constitution

The enumeration in the Constitution of certain rights shall not be construed to deny or disparage others retained by the people.

—Amendment IX to the U.S. Constitution

Copyright © 2000, by Charles N. Hill

All rights reserved.
No part of this book may be reproduced, stored in a retrieval system, or transmitted by any means, electronic, mechanical, photocopying, recording, or otherwise, without written permission from the author.

ISBN: 1-58820-363-8

Second Edition

ABOUT THE BOOK

Fix on the Rising Sun is more than a tale of piracy and murder. It is, as well, a "bill of indictment" which may, ultimately, close the books on one of the darker events in aviation history: the disappearance, on July 29, 1938, of Pan American Airways' trans-Pacific flying boat, *Hawaii Clipper*. And, if a proper Epilogue is ever written, it will document the recovery, from a concrete tomb, of her nine crew and six passengers—the *ultimate MIA*'s of the War in the Pacific.

But the *Hawaii Clipper* did not simply "disappear:" she was hi-jacked to Truk Atoll by radical officers of the Imperial Japanese Navy. Her fifteen crew and passengers were murdered and entombed within a slab of wet concrete on Dublon Island, at Truk Atoll, and, quite inexplicably, the United States Government continues to keep this secret *for* the Japanese government—and *from* the American People—as it has, since 1938.

The charge of piracy is well documented and supported by evidence transmitted from the Clipper during her final flight under American colors—westbound, out of Guam. An exhaustive analysis of the flying boat's last five reported positions demonstrates that her flight reports were deceptive and that her falsely reported approach to Manila was in conflict with Pan American Airways' usual flight policies. Most important, the analysis clearly indicates Japanese involvement in piracy by radio-deception and suggests, not only the Clipper's destination, but her true position at the instant that her radio signals ended abruptly—on approach to an Imperial Japanese Navy seaplane base, at Ulithi.

As to the fate of *Hawaii Clipper*'s nine crew and six passengers, the 1964 report, asserting that they had been entombed in a concrete slab, came from two Micronesians, known to have had close ties to the Imperial Navy's Fourth Fleet, as contractors, at Truk, in the late 1930's. (As the foundation of the Fourth Fleet naval hospital, the slab was the later site of numerous medical war crimes.)

And, while this account, as related to the author by Joe Gervais (who interviewed the former Truk contractors in 1964), does not constitute "hard evidence," it surely stands, for all its wealth of detail, as consistent with the documented facts of the Clipper's loss.

Beginning with the Air Safety Board of the CAA, which left the case open in 1938, many have pursued the elusive Clipper, with limited success, and one can only speculate as to their respective reasons for eventually abandoning her. Perhaps it became apparent to each of them that there are forces, not only interested in, but actually intent upon, preventing this story from ever reaching the public eye. Most assuredly, there is reason to believe that *Hawaii Clipper* might unlock many doors to the past and that, beyond those doors, sixty years of history may prove to be little more than a house of cards.

If the victims of flight 103 of Pan Am's *Maid of the Seas*, which was destroyed by a bomb over Lockerbie, Scotland in 1988, are deserving of the national memorial which has been dedicated to them, then, surely, the dead of PAA's *Hawaii Clipper* deserve, at the least, a decent burial. Then, too, it should be apparent that the only valid deterrent to international crime, whether committed by rebels—or by empires—lies not in threats of retribution under the Law, which is so often corrupted by policy, but only in the absolute assurance that Truth will not suffer silently for long, or lie buried, forever.

Although the primary aim of this book is the recovery of the fifteen *Ultimate M.I.A.*'s, Truth, alone, is a goal worthy of pursuit, as well, and if this quest is successful, then the Truth, unveiled at last, may serve both our peoples: the Americans—*and* the Japanese.

—Charles N. Hill

DEDICATION

To an American aviator,

JOE GERVAIS,

without whose pursuit of truth, as I have come to know it—this
book might not have been possible,

and

To a Japanese aviator, the late,

MINORU GENDA,

without whose pursuit of glory, as I have come to believe—this
book might not have been necessary.

AN OPEN LETTER: To the Emperor of Japan

Your Majesty:

Please permit me to offer my respectful congratulations on your Majesty's victory in Washington on June 13, 1994. I refer to your Majesty's diplomatic "beachhead" on the lawn of the White House, fifty years, almost to the day, after U.S. Forces established an invasion beachhead at Saipan, during the Pacific War, the seemingly *unending* war.

As your Majesty surely knows, of all the battles of that war, the only equivalent to the Normandy invasion of June 6, 1944, was the invasion of Saipan, which began at 8:40am, June 15, 1944 (5:40pm, *June 14*, across the date-line, in Washington). It was the turning point of the Pacific War, and, on June 26, Emperor Hirohito declared, "Hell is on us," and accepted Premier Tojo's resignation. On *that day*, in Japan, Saipan was known to be irrecoverable, and the Pacific victory began to slide, inexorably, into American hands.

Fifty years later, following the extensive coverage of the President's commemoration of the Normandy invasion, many Americans expected a similar commemoration of the far more significant Saipan invasion, of only a week or so later, but, as is now evident, the Saipan anniversary was—*overlooked*.

Blinded by the *Rising Sun*, the President simply ignored the anniversary. This, alone, would have been a fine victory, but, he was also induced to extend, to your Majesty, the hospitality of the White House, at his first state dinner, on June 13, 1994, *the eve of the U.S. anniversary of the Saipan invasion*. That the President *bowed* to the Emperor of Japan, *on that date*, desecrating the memory of Americans who lay *forgotten* at Saipan, was, indeed, an Imperial Japanese victory!

Guests were astonished, as your Majesty proposed a Japanese Emperor's first toast to an American President. And yet, considering that, by his "neglect" of this anniversary, the President had effectively apologized for the American victory at Saipan, it was certainly appropriate for your Majesty to raise a glass. Indeed, by his acts, the President may have expressed *official* regret for every American victory of the war, including final victory. Your Majesty's visit was no less victorious, commanding a U.S. diplomatic silence until your Majesty's departure, June 26, the day after the U.S. anniversary of "the beginning of the end" of the Pacific War, at Saipan.

In truth, however, it must be admitted that America owes a debt to your Majesty's father, the late Showa Emperor, whose bold decision to end the Pacific War spared the lives of millions of Americans, both military and civilian, at great risk to his own. Had he not defied his own forces, in the summer of 1945, and exploited the atomic sacrifices as a cause for surrender, many millions would have died in America, *when the late autumn winds blew*, and a million G.I.'s would have died in Japan, *as the winter snows fell*. Yet, his concerns were never for the Americans, of course, but for the Japanese People, his *Hundred Million*, who, most surely, would all have perished, *in the cold spring rains*, had he not "endured the unendurable," and ended the hostilities of the Pacific War.

If the Shoguns of Japan still resent that the Americans, and their own late Emperor, deprived Japan of a national suicide, what is that to Japan—or to your Majesty? Surely, it is demeaning to Japan, to pursue this endless war by extortion, a criminal's weapon. So, if there be truths which the American People should know, but which Japan conceals, in return for American political concessions, then, perhaps, it is time that these truths be told, and that the Pacific War be *truly* ended.

Most Respectfully,

Charles N. Hill

x

ACKNOWLEDGEMENTS

Major Joseph L. Gervais, USAF (ret.), the *sensei* of Amelia Earhart researchers, whose unselfish guidance has sustained so many (including the author, since 1974) in the search for the elusive aviatrix, whose document "package" provided such a firm basis for the analysis of *Hawaii Clipper*'s loss—and whose then-unbelievable warning, regarding the risk in addressing the AERC symposium at Purdue University, in 1989, the author so carelessly disregarded.

John Luttrell, Earhart hunter, good friend and fellow Earhart researcher, who, in 1987, in addition to other support and assistance, provided the copy of Gervais' document "package," the subject of which, until then, had been of passing interest to the author.

The Honorable Jesse Marehalau, Consul, and now Ambassador, to the United States from the Federated States of Micronesia, who aided the author in making contact with Fr. Hezel at Chuuk (Truk), in 1987.

Fr. Francis Hezel, with FSM's Micronesian Seminar, who investigated the tale at Chuuk in 1987 and, although unable to find any living Trukese residents who could confirm the 1938 interments, found sufficient cause to sustain his own interest in the project and to offer continued assistance to the author.

Hope (Wyman) Streeter, daughter of Ted Wyman, *Hawaii Clipper* passenger, who has waited patiently throughout this project and who has so graciously provided the personal profile and travel itinerary of her father, as well as two 1916 snapshots, which establish the long-standing personal link between her parents and Gen. MacArthur's later wartime Chief of Staff, Lt. Gen. Richard Sutherland. Mrs. Streeter has empowered the author to locate and recover her father's remains.

Mary Ann (Walker) Lee, who, with the support of her husband, Travis, recounted long-suppressed memories of her brother, Mark Walker, and of the last evening which she spent with him in San Francisco, before *Hawaii Clipper*'s departure from Alameda. Mrs. Lee has empowered the author to locate and recover her brother's remains.

Captain Robert B. Greenwood, USN (ret.) 1943 Annapolis graduate, naval aviator (and cousin to Mark Walker), whose letter to the U.S. Naval Academy Alumni Association's *Shipmate Magazine*, along with subsequent telephone conversations with the author, in 1987, provided clear evidence of the belief, among PAA flight officers, in 1938, that a Japanese hi-jacking was a real possibility.

Mrs. **Anne O'Sullivan**, daughter of John Jewett, Mrs. **Eleanor Eymann**, his widow and **Jan O'Sullivan**, his grand-daughter; Mrs. **Marjory Wilson** and Mrs. **Lynette Robinson**, Kenneth Kennedy's daughters, and his niece, Mrs. **Roberta Graham**, who provided the transcript of his last letter.

Andy Loree, the Vietnam War Veteran who coined the phrase, *"the ultimate M.I.A.'s."*

Josh Traynor and **Barbara Wiley**, the co-founders of the *Amelia Earhart Research Consortium*, who, most graciously, invited the author to address their symposium of Earhart researchers at Purdue, in 1989; and **Richard Freeman**, a "silent partner" in the AERC, whose well-informed interest in the author's work indicated that the author is on track regarding both the Clipper and Earhart. [Note: experiences of the author, subsequent to the symposium, indicated that the AERC may have served as a national security sting operation, aimed at determining the extent of the apparently sensitive Earhart research.]

Hugo Leuteritz, **Almon Gray**, **Harry Canaday** and many other former PAA personnel of the 1930's, whose recollections of PAA's Pacific operations added depth to the official records and published materials.

Rear Admiral Richard Black, USN (ret.), who provided a great body of material, regarding search operations aboard USCGC *Itasca*, during the Amelia Earhart mission of July and August of 1937.

Captain William L. Sutter, USCG (ret.), who provided personal glimpses into Earhart radio communication aboard USCGC *Itasca* from July 1-2, 1937, including the "shriek," that may not have been Earhart or Noonan.

John L. Johnson, Jr., who, over the years, provided a large body of source material.

Jerome Steigmann, WWII Marine Sergeant (Intelligence) whose true account of Glenn Miller, related to the author 12 years before a "sanitized" version appeared in *Das Bild*, provided confidence that his information on the Clipper could be "taken to the bank."

Dean Magley, who provided copies of several pages from the memoirs of a retired USAF officer, citing an astonishing personal revelation, by JDF General Minoru Genda, that Earhart and Noonan had been held in Japanese custody and executed as spies.

Mike Seckman, Amelia Earhart's cousin, who provided personal details of a 1962 chat with Irene Bolam, which appears to identify Mrs. Bolam not as Earhart, but as a "stalking horse," set out to divert researchers.

David Church, of the Naval Academy Alumni Association's *Shipmate Magazine*, for his kind permission to reprint the letter from Captain Robert Greenwood, regarding the hi-jacking concerns of his first cousin, Mark Walker, First Officer of the ill-fated *Hawaii Clipper*, on Trip 229, July, 1938.

Christine Weideman and **Judith Schiff**, of the Yale University Library, for their kind assistance in providing permission for the author to reprint Richard Sutherland's 1916 Yale University graduation photograph.

Jack Novak, who provided the Signal Corps photo of MacArthur, Sutherland and Umezu, during the surrender ceremonies aboard the *USS Missouri*, in September of 1945.

John Underhill, for photos of the *M-130* model on display at the Treasure Island museum, which provided an aid in preparing the computer drawing of the Martin *M-130*.

Bill Prymak, who provided air travel to Las Vegas following the AERC symposium in '89, enabling the author to meet Joe Gervais.

Tom Prewett, for his generous 1989 offer to underwrite a survey trip to Dublon Island.

Robert Benzon, of the U.S. NTSB, whose interest provided hope that, with the impetus of public concern—and a suitably expanded budget—the case of *Hawaii Clipper* might eventually be reopened by NTSB.

Col. Rollin C. Reineck, USAF (ret.), who propelled me "once more into the breech."

David Wecker, popular columnist for the *Cincinnati Post*, and **Mike Philipps**, the *Post*'s editor, for two rejuvenating columns.

Bob Silverman, **Stan Virden** and all of the Mensans who gave the work-in-progress a fair hearing and the author a warm welcome in December of 1994.

Dave Anthony, Art Soma, Paul Zellhart, and many other friends and fellow workers at the Gradall Company, from 1987 to 1990, whose interest helped to keep things rolling.

Carol, and the boys, who have borne so much, in the quest for the *Ultimate M.I.A.*'s.

BIBLIOGRAPHY

BOOKS

Ayres, Frank, Jr., Ph.D.: *Theory and Problems of Plane and Spherical Trigonometry.* New York: McGraw-Hill Book Company, 1954.

Backus, Jean L.: *Letters from Amelia.* Boston: Beacon Press, 1982.

Bauer, George N, Ph.D. and **Brooke, W. E**: *Plane and Spherical Trigonometry.* Boston: D. C. Heath & Co., 1917.

Cohen, Stan: *Wings to the Orient: Pan American Clipper Planes 1935-1945.* Missoula, MT: Pictorial Histories Publishing Co., 1985.

Daley, Robert: *An American Saga: Juan Trippe and his Pan Am Empire.* New York: Random House, 1980.

Devine, Thomas E. with **Daley, Richard**: *Eyewitness: The Amelia Earhart Incident.* Frederick, CO: Renaissance House, 1987.

Dupuy, Trevor N, **Johnson, Curt**, and **Bongard, David L.**: *The Harper Encyclopedia of Military Biography.* New York: Harper Collins, 1992.

Dwiggins, Don: *Hollywood Pilot.* Garden City, NY: Doubleday and Co., 1967.

Earhart, Amelia: *Last Flight.* G. P. Putnam New York: Harcourt, Brace and Co., 1937.

Goerner, Fred: *The Search for Amelia Earhart.* Garden City, NY: Doubleday and Co., 1966.

Gwynn-Jones, Terry: *Wings Across the Pacific.* New York: Orion Books, 1991.

Jackson, Ronald W.: *China Clipper.* New York: Everest House, 1980.

Jordanoff, Assen: *Through the Overcast: The Weather and the Art of Instrument Flying*: New York: Funk and Wagnalls Company, 1939.

Kaucher, Dorothy: *Wings Over Wake.* San Francisco: John Howell, 1947.

Klass, Joe and **Gervais, Joseph**: *Amelia Earhart Lives: A Trip Through Intrigue to Find America's First Lady of Mystery.* New York: McGraw Hill Book Company, 1970.

Loomis, Vincent V. with **Ethell, Jeffrey L.**: *Amelia Earhart: The Final Story.* New York: Random House, 1985.

Prange, Gordon W.: *God's Samurai: Lead Pilot at Pearl Harbor.* Washington: Brassey's, 1990.

Putnam, George P.: *Soaring Wings.* New York: Harcourt, Brace and Company, 1939.

Roseberry, Cecil: *The Challenging Skies.* Garden City, NY: Doubleday and Co., 1966.

Who Was Who in American History: The Military. Chicago: Marquis Who's Who, Inc., 1975.

MAGAZINE ARTICLES

Gray, Captain Almon A.: USN (ret.). "Sparks Across the Pacific." *Sparks Journal Quarterly* Vol. 6 No. 2: Dec. 20, 1983, p. 16-21.

Greene, Fred: "Flight of China Clipper Riveted World's Attention." *American Philatelist.* October, 1985, pp. 896-903.

Greenwood, Captain Robert: USN (ret.). "Amelia Earhart." *Shipmate Magazine*. Naval Academy Alumni Association, Jan.-Feb., 1987.

Miller, William B.: "Flying the Pacific." *The National Geographic Magazine*, Vol. LXX, No. 6, December, 1936, pp. 665-707.

Noonan, Fred: "A Letter from Fred Noonan to Lieut. Cdr. P. V. H. Weems [May 11, 1935]." *Popular Aviation*. May, 1938, p. 51.

O'Leary, Michael, Editor: "War Crimes Trials" Air Classics, Vol. 30, Number 9, September, 1994, pp. 101-106.

Scammell, Henry: "Pan Am's Pacific." *Air and Space*. September, 1989, pp. 64-73.

Smith, Richard K.: "Transoceanic Aviation Fatalities: 1919-1939." *Journal: American Aviation Historical Society*. Fall, 1990, pp. 182-187.

Unsworth, Michael E.: "Undercover: In Japan's aerial assault on America, balloon-carried bombs produced some strange results." *World War II*. Sept., 1988, p. 8 ff.

DOCUMENTS

Arey, James A., (Pan Am public relations): Letter. To Mrs. Streeter: Refusal to re-open the case of *Hawaii Clipper*, Aug 8, 1980.

Arnold, Donald D, Lieutenant, USAAC: Report. "Statement of Engineering Officer, Hawaiian Air Depot, Luke Field, T.H.," March 25, 1937.

Balfour, Harry J.: Letter. To Leo G. Bellarts: Ref.: Amelia Earhart Communication planning, October 1, 1970.

Collopy, James A.: Letter. To Joe Gervais: Lack of wireless telegraphy skills of Earhart and Noonan, November 1, 1965.

Galten, William L.: Radio Log 0718-1039 (-11.5 zone), USCGC *Itasca*, July 2, 1937, 3 pages "preserved" by Leo Bellarts, CRM.

Hoyt, Robert D., **Salzman, Phil C.**, **Miller, William T**, (members of the Commerce Department Board); **Smith, Col. Sumpter**, **Hardin, Thomas O.** (Advisory members of the Commerce Department Board and sole constituents of the Air Safety Board of the CAA): Report. *Preliminary Report of Investigation of the Disappearance of an Aircraft...on July 29, 1938*.

Kennedy, Kenneth A.: Letter. To Marjory (his wife): Events aboard *Hawaii Clipper* en route to Midway, July 25, 1938.

Priester, Andre A.: Report. "Hawaiian [sic] Clipper." Details of loss of *Hawaii Clipper*; Engineering Department of Pan American Airways, August 2, 1938.

Putnam, G.P.: Letter. To Marvin MacIntyre: Regarding a reference to the "eastern fringe of the Marshalls," July 31, 1937.

Yale University Library. "Richard Kerens Sutherland." *Yale Classbook*, 1916, p. 215.

NEWSPAPERS

The Bronxville [NY] Review-Press, 8/4/38.

The Cincinnati Post, 7/29/38-8/5/38.

The Cincinnati Times-Star, 7/29/38 7/30/38.

The New York Times, 6/3/37 7/1/37-7/19/37 12/13/37-12/30/37 1/12/38-1/14/38 7/30/38-8/4/38 8/7/38-8/9/38 8/18/38.

The Oregon Journal, 7/29/38 10/28/38.

The Oregonian, 7/24/38 7/31/38 8/4/38.

The Times Record News, 4/16/92. (Wichita Falls, TX)

TABLE OF CONTENTS

ABOUT THE BOOK	v
DEDICATION	vii
AN OPEN LETTER: To the Emperor of Japan	ix
ACKNOWLEDGEMENTS	xi
BIBLIOGRAPHY	xiii
TABLE OF CONTENTS	xv
PREFACE	xvi
INTRODUCTION	xxiii
1. LOST SOULS	1
1. THE PASSENGERS	3
2. PASSENGER EDWARD E. WYMAN: The Wyman / Sutherland Connection	9
3. THE FLIGHT CREW	17
4. FIRST OFFICER MARK WALKER: The Greenwood / Lee Reflections	21
5. THE HI-JACKERS	27
2. THE HI-JACKING	33
1. OVERVIEW: The Clipper Hi-jacking of 1938—and the Ultimate M.I.A.'s	35
2. A WESTERN PACIFIC GAME-BOARD: The Philippine Sea Area, 1938	37
3. PAA AIR ROUTES TO THE ORIENT: Bridges over a Sea of Turmoil	39
4. GUAM: PAA Operations in the Area of Guam	41
5. 2 CLIPPER APPROACHES TO CAVITE: 1 as by Sea—1 out of the Blue	43
6. MANILA BAY: PAA Operations in the Manila Area	45
7. PREFLIGHT: PAA *Hawaii Clipper*—Trip 229 / Leg 5, July 29, 1938	47
8. LAST LEG TO OBLIVION: Final Hours of the *Hawaii Clipper*	49
9. MANILA BAY: D-E EXT and E-282T Courses to Manila Bay	51
10. AFTERMATH: A Divided Search and Flight Reports in Doubt	53
11. 3 CLIPPER APPROACHES TO CAVITE: 1 Secure, 1 Efficient—1 Risky	55
12. THE CHANGING FACE OF TRUTH: Analysis Raises More Doubts	57
13. THE NAVIGATOR'S "SUN-LINE": A Graphic "Rising Sun"-Line	59
14. PALAU: The "Rising Sun"-Line to Palau	61
15. A SECOND "RISING SUN"-LINE: The Clipper's Fix on Saipan	63
16. SAIPAN: The "Rising Sun"-Line to Saipan	65
17. A LAST "RISING SUN"-LINE: A Speed-Made-Good Fix on Truk	67
18. TRUK: The "Rising Sun"-Line to Truk	69
19. THE RADIO BEARING *MIRROR*—A Brilliant Radio-Deception Plot	71
20. ADDED DIMENSION: The Shadow of a New Player Crosses the Board	73
21. MALICE THROUGH THE LOOKING-GLASS: The Clipper Comes About—for Truk	75
22. 1 UNWELCOME APPROACH TO CAVITE: On Course—but a Bit too Early	77
23. DEAD-RECKONING A DEADLY GAME: The Perfect Plan—and one Minor Hitch	79
24. THE HI-JACK SEQUENCE: Scenario for an—almost—Perfect Crime	81
25. LINGERING TRACES: As Indelible—as if Cast in Concrete	83
26. TRUK: The Entombment Site on Dublon Island	85
3. SUMMATION	87
1. THE EVIDENCE: A Brief Summary	89
2. THE SCENARIO: Tying the Evidence Together	95
4. APPENDIX	103
1. THE JACKSON / DALEY RESEARCH: Hard-won Pros and Calculated Cons	105
2. THE GERVAIS DOCUMENT "PACKAGE:" Preservation through Dissemination	109
3. EARHART AT MILI: Resolving AE's 157-337 Line of Position—the Reference Point	155
4. THE SAILINGS: Analytical Navigation Mathematics—the Equations and the Data	167
5. AFTERWORD: Opposing Forces—Reflections of the Author	177
6. INDEX	179
7. ABOUT THE AUTHOR	187

Dear Charles,   May 27/96

    Your many years of research, astute analysis, and data presented in " Fix on the Rising Sun " is a tremendous achievment. This book will certainly cause a major readjustment on the final disposition of the Hawaii Clipper and its flight crew and passengers. This work far surpasses the effort of R.W. Jackson- his China Clipper book published in 1980 revealing Japanese Naval Officers as the catalyst for hijacking the Hawaii Clipper.

    Enclosed is some info on Capt Terletcky and Earhart, on July 2/37 both are flying over the Pacific on the same day, one is heading east and the other heading west. Whats intering, or call it the long arm of coincidence, that within one year both with encounter an intervention by the Japanese.

    I have also included the War Crimes report- Truk during WWII in the area of the hospital and atrocities of American POW's by hospital corpsmen and doctors.

    In 1964 while diving in the area of the Northeast Pass of Truk Atoll I located a B-24 Liberator shot down by the Japanese while on a photo recon mission. I removed both ID plates from the number 3 and 4 engines and both props were in the feather postion. The crew was captured after ditching in a shallow reef area and taken to hospital area on Dublon. I still retain the ID plates and the delivery date of this aircraft was July 14,1943.

    I have written to Prymak & Col. Reineck to order your book "ASAP "., again many thanks for the dedication in your book and good luck with a hardcopy edition soon.

Joe Gervais
Joe Gervais
Major, USAF, Retd
Command Pilot.

Letter from Joe Gervais   To the author, from noted Earhart researcher, Joseph L. Gervais, with his gracious appraisal of the limited First Edition of *Fix on the Rising Sun*.

PREFACE

I. The most reliable evidence for solving an old mystery is that which was contemporary to the original events, and which has lain unseen and undisturbed since those events. Less effective are those few bits and pieces that frequently emerge from the human memory—and conscience—as events begin to slip into history. Yet often, even these fragile shards not only provide the leads by which hard contemporary evidence may be recovered, but also constitute, in themselves, a necessary part of the preponderance of evidence which is often required to fill a gap in official history. But in some especially sensitive cases, those which involved (and may still involve) international politics, the most viable evidence may be contemporary case information, which was disseminated reliably, for whatever reasons, before any official silence could be imposed.

Regarding the hi-jacking of *Hawaii Clipper*, the best contemporary evidence, presently extant, is just such data, received from the Clipper in the two hours before she suddenly ceased radio transmissions. This data entails her last five reported positions, which were recorded only in a Pan American Airways *Engineering Report*, released to a limited distribution on August 2, 1938 (five days after the hi-jacking) by PAA Chief Engineer Andre A. Priester and designated, in the Pan American Foundation (PAF) archives, as files 10.10.00, 30.20.02 and 50.02.01.

Not long after the "demise" (or demises) of Pan Am, the PAA files were transferred to the Richter Library of the University of Miami, under a joint agreement with the Pan Am Historical Foundation, which had provided funds for the purchase of these files during Pan Am's bankruptcy period. The author has been informed that a number of files were lost or destroyed in transit.

Of the five reported positions, contained in Priester's *Engineering Report*, only the last was included in the report of the Air Safety Board (of the Civil Aeronautics Authority), which was released on November 18, 1938, just about sixteen weeks after the hi-jacking. Inasmuch as the ASB felt constrained to leave the case open, this was entitled a *Preliminary Report* and was concluded, "with the thought that additional evidence may yet be discovered and the investigation completed at that time."

These five positions were the subject of an extensive analysis by the author, which not only provided a foundation for the present work, but, in the sense of "new evidence," prompted a formal request to the National Transportation Safety Board, dated July 26, 1994, to examine the author's analysis and to reopen and complete the investigation. That request was denied.

It may be argued that only five reported flight positions, along with a small number of known geographical positions, constitute a rather slim body of evidence by which: to analyze the flight; to determine that the hi-jacking had, in fact, taken place; to assign culpability for this act of piracy; to establish the aircraft's true destination and to make all of this clear to most readers, who have little knowledge or interest in air navigation. Moreover, it may appear doubtful that a hi-jacked commercial air crew, facing certain and impending death, could have embedded so much information in so few reports—and in so little time (about five hours). In fact, the Clipper crews may have been better prepared for piracy than previously thought, And, as will be seen, the limited data proved *entirely sufficient* to the analytical task, and, supported by additional information, makes a strong case for the fact of a hi-jacking.

And yet, while the flight analysis required a great deal of time and effort, it was the more difficult problem of *presenting* this evidence which demanded 28 mathematically-precise, computer-generated charts and eleven years of design and development. The chart series provides a step-by-step presentation of the analysis, augmented by accompanying text, so that most readers should be able to follow the story without feeling intimidated by the complex mathematics of navigation.

The somewhat large size of the book, itself, was mandated by a need to provide large-scale charts, with the accompanying texts located conveniently alongside the charts. A large, twelve-point font was chosen so as to accommodate older readers, and the wide page margins were specified to enable those with prior navigational experience to make notes on the charts. For armchair navigators, an extensive Appendix has been provided, complete with positional data, the "sailings" (i.e., the navigational equations) which were employed and copies of the key documents upon which the analysis was based. In short, it took over eleven years to produce a report which may be understood by almost anyone who can read a map, but which can also stand up to the critical judgment of the most experienced air navigation specialist.

Even given that I may have succeeded in all this, it may still be argued that the "True" positions shown on chart C-0, that is, those leading from the Clipper's noon position to Ulithi, are not factual, but only a matter of speculation. But this is not at all the case. As will be shown on C-12, the dead-reckoned "True" positions are the only mathematically possible alternatives to the reported "False" positions, given *Hawaii Clipper*'s reported flight parameters and the last Makati RDF bearing, both as reported by *Hawaii Clipper* in her last radio report. That is, if the Clipper were not on her reported route to Cavite, at the time that her radio transmissions ceased (which is evident), then she can be dead-reckoned to *nowhere else* but on course to, and fifteen minutes away from, an Imperial Japanese Navy seaplane landing area in the lagoon at Ulithi Atoll.

II. This is not the first investigation of the *Hawaii Clipper*'s disappearance and loss. It is, actually, the sixth known investigation, and the third in a contiguous series, which began about twenty years ago.

The first, of course, was the investigation that took place as part of the search for the aircraft in July and early August of 1938. As will be shown, there were several puzzling developments during the search, involving a false oil slick, an overflight of the eastern Philippines and a very curious interchange between Japan and the U.S., conducted on diplomatic and press levels, and culminating in a *confirmation of the fifteen fatalities*, by the Japanese This frank confirmation, which had originated in Tokyo (and was reportedly released, for a brief time, in the U.S.), was inexplicably squelched by the *New York Times*. Although the results of this search-related investigation were inconclusive, the question of Japanese involvement was raised in the press, both directly and indirectly.

The second investigation was conducted in the fall of 1938 by a Commerce Department Board of Inquiry, which was transferred, during the period of its investigation, to the jurisdiction of the Air Safety Board of the then-new Civil Aeronautics Authority. The first Board convened in Alameda on August 18, 1938, and immediately determined the inadvisability of holding public hearings. By permission of the Secretary of Commerce, who maintained political control, the Board continued its investigation in secrecy and, on November 18, 1938, the original Board released its *Preliminary Report* under the aegis of the Air Safety Board of the CAA.

The report made an undisguised and poorly substantiated effort to direct attention away from the sea area of the divided search and into the jungles of the Philippine Islands. However, to its credit, the Board reached no conclusive determination as to the fate of the *Hawaii Clipper*, and left the case "open." On the other hand, the report revealed none of the several "conjectures," which had been proposed (and which it had simply refused to consider) and neglected to report the most important details of the Clipper's last flight, that is, four out of the five final reported positions, the fifth position having already been widely reported in the press. In this case, the fifth and final position was among those important pieces of contemporary evidence which are often (and in this case, *were*) disseminated, before a lid of secrecy could be imposed.

The third investigation apparently occurred under security wraps in Japan, immediately following the Pacific War, and is known to have taken place only through two separate verbal reports, made to civilians in 1945 and 1946. One report was made to Juan Trippe, president of Pan American Airways, by Admiral John H. Towers, who had retired in 1945 and then accepted a position as a vice president of PAA. Towers' report to Trippe was cited by Robert Daley in *An American Saga: Juan Trippe and his Pan Am Empire*, and described an *Hawaii Clipper* magneto housing serial number, which, presumably during a reverse engineering process, had been molded into the magneto housing of every Mitsubishi A6M2, the original "Reisen" (*Zero*) fighter. In addition to telling Daley, Juan Trippe recounted Towers' report to others at PAA, one of whom, Harvey Katz, wrote a memo on the subject, which may still exist within PAF file 10.10.00.

The other report originated with Lt. General Richard K. Sutherland, MacArthur's Chief of Staff. According to Hope Streeter (the daughter of *Hawaii Clipper* passenger, Ted Wyman), her mother, Mrs. Edna Wyman, had informed her, shortly after the Pacific War, that a report had come to her, from a "General Sutherland" (whether directly or indirectly is unknown), that aircraft engines, known to have originated with the *Hawaii Clipper*, had been recovered, in Japan, after the war. As will be shown, there is good reason to believe that General Sutherland did, indeed, contact Edna Wyman directly in this regard, even though this would have entailed a serious security breach. Although nothing else is known about this secret third investigation, these two singular reports, both related to the engines, appear to be genuine. Related to this investigation, but by no means contemporary, in the sense applied to the reports above, is the report, as told to the author in the late 1990's, of the Clipper's having been seen, virtually intact, in 1945, in Yokosuka, Japan. The author's informant was a wartime Marine Intelligence sergeant whose veracity is known to the author.

The fourth investigation surfaced in 1980, with the publication of Ronald Jackson's fine book, *China Clipper*, which, despite its title, detailed not only the final flight of the *Hawaii Clipper* and its passengers and crew, but meticulously outlined the history of PAA's Pacific enterprise, and of the many Japanese attempts to sabotage the Clipper operations. Although presently out of print, Jackson's book is still available in many Public Libraries and is recommended highly as a companion to the present analysis.

Jackson relied almost entirely upon evidence contemporary to the period that preceded the hi-jacking and founded his own verdict of piracy upon Federal records of Japanese involvement in covert acts of espionage and sabotage against PAA. Of great interest in Jackon's text is his report of the rumors of an impending Japanese action against the PAA Clippers, which surfaced at the Commerce Department in late June of 1938, similar to warnings which preceded the bombing of Pan Am's *Maid of the Seas*, over Scotland, in 1988, the infamous Pan Am 103 tragedy.

As will be shown, Washington insiders had good reason to be fearful of Japanese action against any Clipper that was scheduled for Trip 229, especially considering who, and what, were en route to China.

The fifth investigation, in the summer of 1980, was rather short lived, but brought together two bodies of evidence that were contemporary to 1938. In 1964, Joe Gervais, a retired USAF officer and dedicated Earhart hunter, had been permitted to visit Truk Atoll, where he had hoped to find a trace of Earhart's Lockheed *Electra*. Instead, he came away with a tragic tale, told by two Trukese eyewitnesses (and participants, as well) to fifteen Americans who, as they said, had been sealed in a slab of concrete on Dublon Island in the summer of 1938. Unhappily, as it turned out later, Joe also came away with a number of pictures of a wrecked aircraft lying in the lagoon at Truk.

Sixteen years later, after Jackson's book was released, Joe Gervais immediately made the connection between the fifteen passengers and crew of *Hawaii Clipper* and the fifteen Americans who, as he had been told, had been entombed on Dublon Island in 1938. Digging out the photos and records of his 1964 trip, Joe contacted the press and, on July 6, 1980, the *New York Times* and the *Los Angeles Times* carried an Associated Press story by Patrick Arnold, recounting Joe's realization of a connection between the Clipper crew and the Americans entombed on Dublon Island. Unfortunately, Joe also suggested that the wreckage, which he had photographed at Truk, might be the Clipper.

Pan Am's answer to Hope Streeter's request, that Joe Gervais' claim be investigated, was quite negative and rested primarily on the wreckage photographs. The photos showed that the wreckage could be proven *not* to be that of an *M-130* flying boat, and so, PAA's Director of Corporate Public Relations, James A. Arey, was able to reject the burial report, as well. Gervais' response was to gather together all of the photographs and documents relating to his trip to Truk in 1964 and to his brief investigation in 1980, and to distribute them to a number of Amelia Earhart researchers and associates, including John Luttrell of Atlanta.

The sixth, and the latest, investigation began almost immediately after John Luttrell sent the author a copy of the Gervais "package," in 1987. John and I had cooperated in our own Earhart investigations for a number of years, sharing information, bouncing ideas off of each other, and, late in 1986, double-teaming an unlucky lady at the Treasury Department, who was suddenly transferred after admitting, in desperation, that a secret Earhart file was still in the possession of Treasury. Although Luttrell continued to pursue his Washington contacts, I regarded the Treasury incident as a brick wall, with regard to Earhart. But, in 1987, I perceived that the Clipper hi-jacking, if true, might provide a "back door" into the Earhart case. However, the story of the *Hawaii Clipper* soon took on a life of its own, and has become for me, over the past years, both a consuming passion and a serious obligation.

Jackson had worked from contemporary evidence and private testimony to provide a vivid picture of the passengers and crew and of the threat to Clipper operations posed by Japanese espionage and sabotage. Gervais had not only been a witness to the important contemporary testimony provided by the two eyewitnesses to, and participants in, the Truk entombment, but had also bundled his travel documents with PAA's *Engineering Report* and the ASB's *Preliminary Report*, ensuring the survival of both documents and the connection to the Truk entombment. For my part, I have provided a math analysis of the Clipper's reported flight data, also contemporary to the period, which, in and of itself, clearly indicts radical elements of the Imperial Japanese Navy—for piracy.

With some copyrighted exceptions, I have reproduced Gervais' document "package," as it was provided to me by John F. Luttrell and have also included a brief compendium of Jackson's vital evidential citations, as well as important details of the hi-jacking, as told to Robert Daley by PAA's Juan Trippe.

Finally, I have also included reports from a few surviving relatives of the Clipper's crew and passengers. Among these are Captain Robert Greenwood, USN (ret.) who, as the fifteen-year-old first cousin of First Officer Mark Walker, had discussed the possibility of a Japanese hi-jacking with Walker during the PAA pilot's hometown visit to Wichita Falls, Texas, three weeks before the Clipper hi-jacking. Walker's sister, Mary Ann Lee, was kind enough to share the memory of her brother's final evening in Oakland, as well as her recollections of the aftermath of the tragedy. Among the most important, with regard to contemporary recollections, are the snapshots provided by Mrs. Hope Streeter, daughter of passenger Ted Wyman, which substantiate the report that her mother had been informed, in 1946, by General Richard Sutherland, of the post-war recovery of the Clipper engines—in Japan.

Here, for the first time, all six investigations have finally converged, in a limited fashion, admittedly, but with such mutual agreement as to provide an argument for a seventh and final investigation; that is, the recovery of the remains of the Clipper's passengers and crew, the ultimate MIA's of the Pacific War.

III. *The seventh* investigation, by all rights, should prove to be the simplest of all. There would be no need for months of research, no poring over endless files and records and no complex mathematical analyses. A survey of the slab should require only a few days' work, using short-range microwave survey equipment, and, given that fifteen human-sized "voids" could be located within the slab, using just such a radar unit, a forensic examination of core samples of some or all of those "voids" should require no more than a week to complete. Upon determining the existence of human remains within the slab, the permission to exhume should be readily available from both the Federated States of Micronesia and the Truk (now Chuuk) State government. The grant of such permission is likely, considering that the Japanese have already been granted such permission, with regard to Japanese remains that have been recovered on the islands of Truk.

Cutting the concrete slab into blocks would provide jobs for the Trukese of the area, and international attention would surely enhance the important tourist trade. Careful removal of concrete around the faces of the remains (generally located near positive imprints on the bottom surface of the blocks) would provide death masks by which to simplify the preliminary identification of the remains (from photos). The removal of the remains, from the blocks, could be set up, in a few weeks, implemented with power saws, and ultimately completed with hand tools. The remains would then be returned to the United States for proper burial and a fitting memorial—no less deserved by these fifteen men, than a national memorial at Arlington is deserved by the victims of Pan Am 103.

Ironically, the greatest obstacle to the seventh investigation will originate, not with Japan, for whose tourists the slab has long been popular (incomprehensibly popular, unless many arrive already knowing what is contained within the slab, a result, perhaps, of the 1980 publicity). The major obstacles will originate in Washington, D.C., where the U.S. Government continues to suppress the truth about the hi-jacking.

That the U.S. suppressed the truth in 1938 is understandable: The *Hawaii Clipper* was hi-jacked by a radical element of the Imperial Japanese Navy, acting independently, even as the U.S. gunboat, *Panay*, was attacked by this element in 1937. In the case of the *Panay*, sunk in plain view in a war zone, the Japanese government's immediate apology and promised reparations were appropriate. For piracy and murder aboard the Clipper, however, neither apologies nor reparations would have been sufficient to quell a loud American demand for vengeance and blood. In 1938, however, the U.S. Navy was only a paper tiger in the western Pacific.

Moreover, as it now appears that three of the Clipper passengers were en route to China to help create a secret, U.S.-funded air combat squadron, similar to the later (1941) A.V.G., (better known as the *Flying Tigers*), the U.S. Government had a more important reason to prevent the American people from learning the truth about the Clipper: it would have been one thing for the American people, themselves, to abandon isolationism in a just cause, but it would have been quite another for them to learn that the U.S. Government was secretly involving them in a foreign war. It should be noted, with regard to the *Flying Tigers* of the American Volunteer Group (AVG), that it was not until 1992 that the U.S. Government finally recognized the official military status of these pilots and admitted to having financed their operations.

Since the end of the Pacific War, however, the American government has continued to maintain the silence surrounding the *Hawaii Clipper*. The private and separate revelations of Admiral Towers and General Sutherland, regarding the recovery of the Clipper's engines in Japan, clearly indicate that her fate was known, immediately following the war. Yet, this fact was not disclosed to the American people, but only to a few private individuals having a personal interest in the Clipper and a personal connection to these two senior wartime officers, indicating that an unpopular cover-up was in place. It is worth noting, with regard to Admiral Towers' story of a Clipper serial number cast into the *Zero* magnetos, that no *Zero*, extant today, has an original magneto installed, including the engine of the *Zero* undergoing restoration at the Air Museum at Wright Patterson AFB, in Dayton, Ohio. Moreover, shortly after Joe Gervais, an Amelia Earhart researcher, had been told of the concrete slab burials, U.S.-administered Truk (where very few people should ever have heard of Earhart), instituted the "Amelia Earhart Law," which forbade all private searches for human remains. This clearly indicates that U.S. officials felt constrained to discourage any follow-up on the sensitive entombment report, which had surely been provided to the U.S. long before Gervais' trip in 1964.

The apparent U.S. opposition to efforts to recover, or even to reveal the passengers and crew of *Hawaii Clipper*, seems especially strange in light of the war crimes trials, held at Guam after the war, in which more than a hundred Japanese were found guilty of war crimes committed at Truk, some of these at the very site where the men from the Clipper were sealed in concrete. In a summation of the crimes, note was even taken of "thirteen white victims whose nationality could not be established." If one considers that passenger Wah Sun Choy was Chinese and that Third Officer Jose Sauceda was Hispanic, that certainly leaves thirteen whites, perhaps an oblique reference to the Clipper.

But why? Why has the U.S. felt compelled to protect a very few radical Japanese naval aviators from a just charge of piracy and murder? One may, as well, ask why the U.S. has supported Japan, at such a cost to the American people, in so many issues, over the past fifty years. The answer may actually involve simple blackmail: a Japanese silence regarding certain mind-blowing events of the Pacific War, in return for certain U.S. concessions on policy and trade: that is, U.S. political disgrace, never revealed, in return for Japanese victory spoils, never won.

—Charles N. Hill

INTRODUCTION

Zoom Out: If one were to poll the people of the United States as to *three* Union issues of the American Civil War, it is most probable that the primary responses would have to do, first, with the issue of slavery and, second, with the issue of preservation of the Union. Most people would be hard-pressed to come up with a third issue of any kind, and few, indeed, would mention the issue of ensuring a cost-effective supply of raw cotton for the northern textile mills. Yet, that issue became the foundation, not only of the war, but also of the other two issues: Young men, toiling on family farms, may have found the anti-slavery issue sufficiently motivating as to warrant their departure for the battlefields. Mature men, with homes and families, may well have found the issue of the preservation of the Union sufficiently motivating as to warrant a risk of sacrificing personal unions. But it was the *men of governance*, North and South, who found the issue of cotton to be sufficiently motivating as to warrant a civil war and the consignment of a half million men to early graves. Yet few of the soldiers would have risked their lives—for *cotton*.

For the two related wars that constituted World War II, there is a similar motivational disparity, which, if a *true* war memorial is to be created, ought to be clearly understood:

With regard to the Pacific War, motivation of the American G.I. was the result of a long history of aggressive Japanese expansion in the Pacific Basin, much of it, admittedly, in response to American expansion. Racist and nationalist propaganda fueled the tension on both sides, while Hollywood's *Charlie Chan* ensured that Americans would not come to hate the Chinese, along with the Japanese. Hostile actions against Americans, in the late '30's (including the suspected Japanese involvement in the Clipper loss), served as an overture to the culmination of all these years of tension: the attack on Pearl Harbor. By then, however, most young Americans had been well primed to confront Japan.

But the War in Europe was another matter. Rumors of Nazi atrocities merely awakened bitter memories of U.S. Government lies and propaganda which had motivated Americans to fight during World War I. Knowing that Americans wanted no part of "Act II" of the Great War, FDR announced, in December of 1941, that no action would be taken against Japan (which had attacked us) until Nazi Germany (which had *not*) had been beaten. Most Americans were outraged, but evolved a kind of *indirect* motivation: as a G.I. told Ernie Pyle, he hoped that Germany could be defeated quickly, so that he could go to the Pacific, "and fight our real enemy, Japan."

Fanatically anti-Communist, Nazi Germany and Imperial Japan had long hoped to crush the Soviet Union. In their Axis alliance, they thought to attack jointly; Germany, through the western Soviet buffer states; and Japan, through Mao's Red buffer in north China. But Germany had, at its back, a strong grass roots Communist movement in France and a high-level Communist infiltration in Britain, which first had to be neutralized. Japan, too, had danger at its back: a nation with both a growing grass roots Communist movement and a government whose policy was directed by a high-level Communist "fifth column": the United States. Oh, yes, the *United States*.

The symmetry dissolved when France and Britain failed to repel the Nazis in 1940, and so, in addition to keeping Japan too busy to invade Siberia, America had to stand up for France and Britain, as well, drawing Nazi fire away from the Red army. With Europe in ruins, the Soviets could feed on its corpse, until Japan's germ warfare finally decimated America, and American G.I.'s exterminated the Japanese in return. That was Moscow's plan and so, in Washington, our own *men of governance*, loyal to the Soviet cause, were motivated, not to crush the Axis so as to end its crimes against humanity, but, rather, to drown the Axis in American blood, so as to ensure the survival of Soviet Communism.

Zoom In: Washington, DC, July 29, 1938: It was not going to be a very good day for aviation at the Department of Commerce, but then, it hadn't been a good year, either. Douglas Corrigan, for example, would soon be returning from Ireland. Having been denied permission to fly the Atlantic, from New York to Ireland, in a tiny, dangerously ill-equipped plane, "Wrong-Way" Corrigan had taken off to return to California—and then turned up in Ireland. Suddenly an Irish-American hero, Corrigan was far beyond any official discipline—or even censure.

Eugene Vidal, at the department's Bureau of Air Commerce, had pumped up so much enthusiasm for flying, especially with regard to his friend Amelia Earhart, that aviation seemed to be almost out of control. But the problems at Commerce were actually moot, since aviation was soon to be handed over to the new Civil Aeronautics Authority, which had been charged with bringing order to the expanding industry. It was not that Vidal had been at fault: he had helped to liberate aviation from the straitjacket of monopoly, and there was bound to be some fallout as aviation blossomed. The problem was that people died whenever a "fallout" occurred in aviation—especially out in the Pacific.

Until the previous July, things had been going so well in the Pacific. After the mad, almost suicidal 1927 Dole Race to Hawaii, in which so many *had* died, the Pacific had slowly yielded to aviation. Kingsford-Smith had reached Brisbane from San Francisco in only three hops, in 1928, and in 1931, Clyde Pangborn had flown non-stop from Japan to Washington State. Of course, Charles Ulm had been lost, without a trace, in December of 1934, perhaps after overshooting Hawaii from Oakland, but less than one month later, in January of 1935, Amelia Earhart had flown the safer reverse course, from Hawaii to Oakland. And, in November of that year, the first of Pan American Airways' Martin *M-130* flying boats, the *China Clipper*, had taken off from Alameda, California, loaded with U.S. mail, and inaugurated a series of over two hundred successful round-trips of trans-Pacific air service to Manila and Hong Kong, a world away. Aviation, it seemed, was beating the odds, out in the Pacific.

Perhaps unavoidably, it had also stirred up international tension, as well. Although the loss of Amelia Earhart, the previous July 2, had had the appearance of being plain bad luck, many Americans were suspicious of something more sinister. In fact, if they had known what was known to more than a few Washington insiders, including her friend, Gene Vidal, they would have been more than merely suspicious.

The public did not know that Earhart had followed her original communications plans precisely, or that a rumored communication breakdown was the result of a tragic change in plans which had been made on the night before her take-off from Lae, New Guinea, or that the changes had been instigated, in her name, during a secret radio "chat" with the U.S. Coast Guard Cutter *Itasca*, by the Lae radio operator Harry Balfour, who had been desperate to join her flight.

They did not know that, aboard the Coast Guard Cutter *Itasca*, at 8:43am on the day of her disappearance, her reported location, "on the line 157-337," had been erroneously interpreted as a radio bearing from Howland Island, due to Earhart's need for secrecy, or that her reported final "flight tactic," that of *running north and south*, was the result of a misunderstanding when, long after 8:55am, a Coast Guard radio operator crammed this vital part of her final broadcast into the 8:43am radio log slot (*Appendix, Section 3*).

They did not know that most of her final radio broadcast, received after 8:55am, had been expunged from the radio logs of the *Itasca* after the Communications Officer, Ensign William Sutter, had heard something "shrieked" from the aircraft, in a voice that seemed not to have been either Earhart's or Noonan's, or that the log falsification would remain hidden until 1975, when the heirs of *Itasca*'s Chief Radioman, Leo Bellarts, sent that long-ago morning's *original* radio log to the National Archives in Washington.

They did not know that on July 5, the PAA Radio Direction Finders on Wake, Midway and Oahu had fixed Earhart's true position in the sea, close to the "eastern fringe" of the Japanese-held Marshall Islands (not down in the Phoenix Islands), or that, soon after, U.S. Navy intelligence officers had ordered PAA secretary, Ellen Belotti, to retrieve all of the copies of these RDF reports—and destroy them. And the public would not know, for well over forty years, that Ellen Belotti had preserved *one* copy of each transcript.

Earhart had been low on gas, or she wouldn't have gone down at sea, but what was she doing near the Japanese Marshalls, some six hundred miles off course? *Lost?* She wasn't at all lost. *Hi-jacked?* That seemed unlikely. True, her *Electra* was quite advanced, but the Japanese had already purchased several from Lockheed. Its Pratt and Whitney Wasp engines were advanced, but the Japanese had been licensed to manufacture engines for P&W. Her radio equipment was nothing special, and Earhart, herself, was neither especially skilled, nor even particularly knowledgeable, as an aviator. In fact, there was little of value aboard the *Electra* except Frederick J. Noonan, the finest air navigator in the world—drunk or sober—and a man who knew Pan American's air operations, policies and aircraft better than most. If he had turned up missing under any other circumstances, Pan American Airways, and, perhaps, the U.S. Navy, would surely have been concerned, from a security standpoint. But lost at sea with a female stunt flier? That should have raised no interest in intelligence circles, because it was not at all unexpected.

And Mark Walker had actually expected it. Earhart had consulted the PAA pilot before she left the West Coast for Miami in June of 1937. As an experienced Pacific flier (who would serve as First Officer in *Hawaii Clipper* on Trip 229), he had been asked by Earhart about Pacific flying conditions, but Walker had been especially negative—and, in fact, brutally frank—about her chances of survival, let alone of success. Walker was concerned that, if her flight were to end in tragedy, public confidence in trans-Pacific aviation, and in the Pan American Clippers, might suffer a considerable setback.

[First Officer Walker was not the only one, aboard the *Hawaii Clipper* on Trip 229, with a connection to Earhart. Fred C. Meier, of the U.S. Department of Agriculture, had been concerned about the transfer of micro-organisms on upper air currents and had contracted with Amelia Earhart to have Fred Noonan collect air samples during her world flight. After New Guinea, of course, the sampling stopped and, during Trip 229 of the *Hawaii Clipper*, Meier was taking the Pacific samples, himself, filling in the blanks left by Earhart's loss.]

If the Japanese had, indeed, wanted to pick Noonan's brain for its wealth of navigation knowledge, then, surely, hi-jacking Earhart's *Electra* would have been the least suspicious way to effect his disappearance. But, hi-jack Amelia Earhart, *Eleanor Roosevelt*'s friend? It seemed so—*melodramatic*. But of course, the Japanese had a flair for melodrama—in life, as in the theater—and her plane did seem to be tail-heavy in that last, dicey take-off from Lae, New Guinea, on July 1, 1937.

But the risk! What if Earhart hadn't waited for Noonan (who held up her flight by flying up to Bulolo on a bender), but had dumped him and had taken Harry Balfour, the Lae radio operator, instead? Then the hi-jacking would have been for nothing. But the plans that Balfour had made with *Itasca* were sound and professional, and a radio operator and technician made far more sense than did a conventional navigator, especially with the weather in doubt and the electrical system subject to overload. But how could a hi-jacker have known that she wouldn't just abandon Noonan straight away and bring in Harry Balfour, who was far better suited to communication and radio navigation—and far more inclined to sobriety? How could a hi-jacker have known just what day she would finally take off? And how could a hi-jacker have known that she would wait so long for Fred Noonan—even sacrificing her plans for a triumphant Fourth of July arrival in Oakland?

For that matter, why *had* she waited so long for Fred Noonan? A bender can go on for weeks—how long had she been prepared to wait? Balfour was a good radio operator, and radio was one of the keys to finding Howland Island. Radio, of course—and an early take-off to ensure arrival before dawn, so that *Itasca*'s searchlights could provide a vivid beacon, visible for about 100 miles. These were the plans of Balfour and *Itasca*, and they were solid plans. Balfour, it seems, could also navigate, but Noonan had no skill as a radio operator and had admitted as much, his first night in Lae: "She can fly—I can navigate—but we both are bum W/T [Wireless Telegraph] operators." Even after reaching Howland Island, would she have needed Noonan to find Hawaii? Waiting for her at Howland Island was Army Air Corps Lieutenant Dan Cooper—and enough 100-Octane gasoline for *two* take-offs. What Lt. Cooper planned to do with her plane, while she slept for twelve hours or so, is anybody's guess, but he could have dead-reckoned her flight to Hawaii just as well as Noonan, and Balfour was even a good bet for that trip, since the Army Air Corps hadn't sent a radio operator to Howland. Why had she waited so long for Noonan? How could a hi-jacker have been assured that Fred Noonan would be aboard? Hi-jacking seemed unlikely, but what *was* Earhart doing, in the Marshalls? And *who* shrieked *what* at 8:55am on July 2, 1937, from Earhart's *Electra*?

[It would not be until August of 1939, that some Washington insiders might understand that Earhart's Japanese hi-jacker hadn't been a hi-jacker at all; that he had been brought aboard because Noonan would certainly have resisted being handed over to the Japanese—by Amelia alone. It would not be until the end of the Pacific War that others might realize that she had defected to Japan as part of an intelligence mission, not for the U.S., of course—and not at all for Japan—but for a social experiment which then stood in grave jeopardy; a social experiment that was so dear to her friend, Eleanor Roosevelt; a social experiment clearly threatened by the Berlin-Tokyo Axis: the social experiment of Soviet Communism. Earhart's mission was to convince Japanese Navy radicals to press for the redirection of Japanese aggression—from the Soviet Union to the U.S. Earhart was, quite possibly, the only disgrace shared jointly by the U.S. and Japan, and the legacy of her mission was the Cold War.]

But Earhart's loss was only the beginning of a new round of aviation crises in the Pacific. Six months later, on December 12, 1937, a U.S. gunboat, the *Panay*, was sunk, on the Yangtze River, by Japanese naval aircraft, supposedly acting without orders. And on January 11, 1938, after reporting an engine oil leak on his *Samoan Clipper* and aborting a mail flight from American Samoa to New Zealand, PAA Captain Ed Musick, who had already jettisoned fuel from this type of aircraft (a Sikorsky *S-42*), radioed that he planned to jettison all of the excess gasoline aboard, before returning to the "tea-cup" (as he had called it) which served as a harbor at Pago Pago, American Samoa. After that last message, nothing more was ever heard.

A few hours later, word reached the U.S. Navy, at Tutuila, that a group of Samoans had seen smoke rising from a fire blazing on the surface of the sea, some fourteen miles northwest of Pago Pago. The Navy quickly dispatched the tender *Avocet* to the area and soon confirmed that *Samoan Clipper* had, indeed, come to grief—with no survivors. This was terrible news, of course, but, at least, the tragedy had not involved any of the big Martin *M-130*'s, PAA's trans-Pacific Clippers, which were engaged in regular air service between San Francisco and Manila, with an extension to Hong Kong. *Samoan Clipper*, a Sikorsky *S-42B*, had recently been withdrawn from service as *Hong Kong Clipper* (on the Manila to Hong Kong leg), and, for service on the New Zealand mail route, had been fitted with long range fuel tanks, as had her predecessor, the *S-42* survey boat, *Pan American Clipper*.

Ten months before, on the first leg of the New Zealand survey flight, March 17, 1937, trailing behind Earhart's *Electra* and *Hawaii Clipper*, en route to Honolulu, Musick had lost an engine on *Pan American Clipper* and had begun jettisoning excess fuel. Gasoline soon began to condense within the hull and Musick was forced to finish the flight with the main switches off and all windows open. PAA conducted tests, which indicated that fuel jettisoned from flush drain-holes flowed forward under the wing, then back over the upper surface—toward the engine exhausts. It was therefore announced, in 1938, that *Samoan Clipper* had exploded in flight due to a fuel explosion, resulting from this quirk in the fuel jettisoning procedure. But, of course, the only reason that PAA was able to identify this quirk was that Musick had been able to *survive* the procedure in 1937.

Still, it seemed a convenient explanation, but one that ignored several key points. *First*, although Samoan witnesses had accurately identified the site of the tragedy, their claim that the aircraft was burning on the sea was simply disregarded: the Navy believed that it had exploded in the air. *Second*, much of the partially charred debris was coated with what appeared to be powdered aluminum, which was far more consistent with a high explosive, than with gasoline, as a source of the explosion. [According to Jackson, PAA later informed the FBI that *a bomb within the hull had destroyed Samoan Clipper!*] And *third*, of course, Musick had already dealt, quite successfully, with the dangers of dumping gas from a Sikorsky *S-42*.

But the real mystery, regardless of what may have caused the explosion, was the absence of any trace of human remains. Although the charred debris, log sheets and clothing found in the oil slick at the site were shown to have originated with *Samoan Clipper*, no human remains, or even partial human remains, were found at the site—*none*. No one who has been at the site of an aircraft disaster at sea—warm water or cold, sharks or no—has been spared the sight of human body parts.

Was it so unlikely that hi-jackers had forced the flying boat to land on the sea, then disembarked the crew into a small boat, a submarine, or even a Kawanishi *H6K* flying boat, then set the fuse on a high explosive package and slipped away with the rest? It would have been reasonable, indeed, for the Japanese: these *S-42B* flights represented an attempt to create a *southern naval air route* to the Philippines, by way of New Zealand and Australia, and Ed Musick represented a more up-to-date source of information than Noonan, even if Fred had talked. Hi-jacking was only a possibility, but it was certainly a chilling one for PAA and Commerce to face.

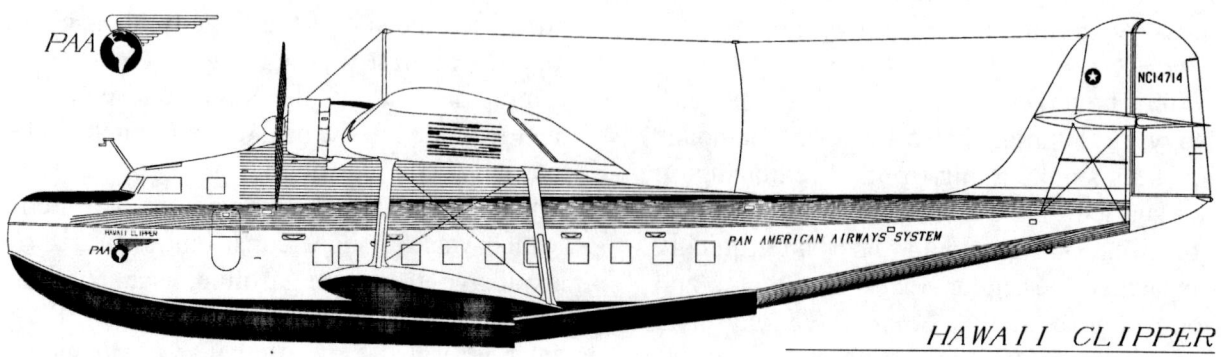

HAWAII CLIPPER

For Pan American Airways' air crews, regardless of what they knew or what they only suspected, *Samoan Clipper*'s loss was a clear sign to begin considering what action could be taken if *they* were ever hi-jacked in flight, and to plan what *they* might do, if only to "get the word out" as to why they would never be coming home again.

Then, in June of 1938, rumors had begun to surface at the U.S. Commerce Department, dark rumors of an impending, but undefined, Japanese action against the Pacific Clippers.

But had Naval Intelligence actually learned of such a Japanese plot? Or had an attack simply been anticipated, due to a specific cargo or passengers that a particular Clipper would soon be carrying? PAA crews had good reason, but little time, to think ahead.

Mark Walker had thought ahead: early in July of 1938, Walker had gone home to Wichita Falls, Texas, for a long overdue visit, before leaving the Pacific Clippers and taking command of a new Boeing *314* flying boat—to inaugurate PAA's prestigious trans-Atlantic run. During his visit he spent some time with his younger cousin, fifteen-year-old Bob Greenwood, who, understandably, admired the dashing Clipper pilot. Bob had, of course, asked his cousin what dangers he had faced in flying the Pacific, and Mark had replied that he had only one real concern for the Pacific Clipper runs—that, without much trouble, *a lone Japanese, armed with a gun, could take over a Clipper in flight*!

As it turned out, the delivery of the new *314* boat was delayed, and, when another PAA pilot fell ill, Walker, who needed just a few more hours in the air to qualify for a Clipper command, had been ordered to fill in as First Officer of the *Hawaii Clipper*, on Trip 229.

All in all, things had not gone entirely well for Pacific aviation during the previous year, and then suddenly, fifteen more—*lost!*

1. LOST SOULS

1. THE PASSENGERS

Choy Wah Sun, 38,
Jersey City, New Jersey:

Mr. Choy was a successful restaurateur, with a Manhattan tearoom and two restaurants in Jersey City, and it is sadly ironic that this aviation enthusiast's best known restaurant was named for the *China Clipper*. Born and raised in America, Wah Sun had grown up accustomed to using the eastern style, that of family-name-first, only within the Chinese community. He had also adopted a western given name, phonetically similar to his own Chinese name, and, among his American associates, Choy Wah Sun preferred to be known as "Watson" Choy. As with many western-born Chinese, however, he still held strong ties to China and reported that he was making the trip to visit a younger brother, Frank, who was serving in the Chinese Air Force. Other reports, however, indicate that Watson Choy had raised a great deal of money for China war relief and that he was carrying some three million dollars to the Kuomintang government in Chungking. He must have been quite thrilled to anticipate crossing the Pacific by air, although he may have been disappointed that he would not be flying aboard the *China Clipper*. That name, itself, was magic, and even Major French had voiced a similar disappointment.

During the first two legs of the flight, from Alameda to Pearl Harbor and, after a layover of one night, from Pearl Harbor to Midway, he had struck up an acquaintance with PAA traffic manager, Kenneth Kennedy, a fellow passenger. Kennedy noted, in a letter written during the flight to Midway, that when they passed through the rough air inside clouds, "our Chinese friend is a little concerned." He noted that Mr. Choy owned a restaurant in New Jersey, adding that, "yes, he calls it Joisey." Choy apparently told Kennedy that he had never been to China, and, despite all that has since been learned about Choy's purpose in traveling to the Orient, Kennedy wrote that, "He says he is going to China for fun. He has a funny idea of fun."

Howard C. French, Major, USAACR
Portland, Oregon:

Although mistakenly identified in the initial news releases as a building contractor from Tacoma, Washington, Howard French was, in fact, a successful auto dealer in Portland. A bachelor, who lived with his mother and his sister, Irene, and who, presumably, could well afford the expensive air passage to Asia; French's reasons for this trip have always appeared a bit vague. On July 29, when news of the Clipper's loss first became known, the *Oregon Journal* put some spin on one of French's earlier statements and declared that "his trip was intended to be that of the first Portlander to make the entire journey to Hong Kong by air." And then, on August 4, when hope had begun to fade, the *Oregonian* noted only that French had been "on a vacation air trip to Hong Kong." Yet, before the trip, Major French had made it quite clear that he was en route to *China*.

Although Major French never made public his reasons for the trip, he had told a reporter from the *Oregonian*, on July 23, just before his departure, "I hope to get up to Canton while I'm there. I want to be in Canton when the Japs pull another raid on the town. I want to see how the bombs drop now. It's been quite a while since I was in the aviation corps during the world war. I saw plenty of them drop then, but maybe the styles have changed."

For Major Howard C. French, senior reserve air officer in the Pacific Northwest and the active commander of the 321st Observation Squadron (an Army Air Corps reserve flying unit, based at Pearson Field in Vancouver, Washington), interest in Japanese bombing techniques was hardly a matter of curiosity. Recognized by the *Oregon Journal* as "one of the foremost air authorities in this part of the country," French had served as Oregon's aeronautical inspector and was both vice president of, and technical advisor to, the Aero Club of Oregon. Before World War I, as a young soldier of the Washington State National Guard, French had left college to serve with American forces on the Mexican border, and, when the U.S. entered the First World War, he saw action at the front, as an infantry Second Lieutenant. After a transfer to the U.S. Army Air Corps, French again saw action, this time as a pursuit pilot with the 50th Squadron, where he twice survived being shot down in flames, to become an ace and a First Lieutenant, and to receive the *Croix de Guerre* from a grateful ally.

In 1938, Major French may well have seen the war in China as an unavoidable prelude to American intervention, even as he had so viewed the Mexican border trouble and the early years of trench war in Europe, prior to the U.S. entry into WWI, in 1917. He surely realized, in 1938, that he and the men of his squadron could soon be in the thick of war in Asia, and that he ought to steal a march on fate, as he had done in Mexico, in 1916.

Whatever reasons Major French may have had, for traveling to China, he seems to have been caught off guard by all the unexpected local publicity which attended his departure from Portland. On Friday evening, July 22, the night before his United Airlines flight from Portland to San Francisco, to connect with the *Hawaii Clipper*, he had been given a surprise party by a group of friends "from all around the northwest." Yet, according to the *Oregonian*, French, who was never a man to choke under pressure, "was not only surprised but delighted beyond words to express his feelings." It appears, however, that Major French may actually have been stunned into silence by all the *unwanted* attention attending his departure for China.

The next day, while picking up his travelers' checks at the local American Express office, French was approached by a Portland police officer, who presented a warrant and then informed French that he was under arrest! "French," as the *Oregonian* later reported, "was startled. He didn't stop to read the document," and, as he was hustled outside to the paddy wagon, French lost his temper and protested that he hadn't "done a damn thing to be arrested for." It had all been a joke, of course: Howard French was a mover and shaker in Portland and enjoyed the level of friends who could involve even the Portland Police in a practical joke. In the letter to his wife, Kennedy referred to him as "Major French who is Irish, a politician who looks like a politician."

French may have been so preoccupied with the nature of his trip to China, that he hadn't seen the predictable "arrest" gag coming. After the high speed, "Keystone Cops" ride to the airport in the Portland paddy wagon, with all his pals packed inside and the siren screaming all the way, French seems to have surrendered to the inevitable and to have given up on any hope of secrecy, or even of discretion, regarding his trip to China. In an interview at the Swan Island Airport, French spoke openly—and mostly of war. He spoke of his own prior experience in World War I, and he spoke of the then-present air war in China and of the brutal Japanese bombing of Canton and its civilian population. War was clearly on his mind, and it may have been that he then regretted having confided in his friend, Neil Bertrandias, regarding what he, and a few others, perhaps, had hoped would be a discreet departure for the orient.

Major French knew his travel schedule well: he noted that "it will take a week, to the very hour, to reach Hong Kong," adding that "I expect to be over there at least ten days, that is, ten days in Manila, Hong Kong and Canton." With PAA's one-night layover in Manila, French was to be a total of nine days in Hong Kong and China, before returning through Manila. He added, as if in answer to a question, "Oh, yes, I'm coming home in about a month; I don't think I want any job in the orient right now." And yet, such a question would have been justified: Howard French, a Major in the Air Corps reserves, would have made an ideal advisor for an air combat fighter squadron: not only was he qualified for such a job, he was, technically, at least, still a "private citizen."

But time had run out. Not only was *Hawaii Clipper* about to become the first airliner in history to be hi-jacked, with the loss of all on board, but barely a month later, Japanese pilots would again make history. In August of 1938, while on a scheduled flight from Hong Kong to Chungking, the *Kweilin*, a *DC-2* operated by PAA's China subsidiary, CNAC, would become the first commercial airliner to be attacked and *shot down*. The loss of the *Kweilin* and its V.I.P. passengers, to carrier pilots of the Imperial Japanese Navy, along with earlier advances made by the Japanese Army, had severely disrupted connection schedules for the PAA Clippers.

Kenneth A. Kennedy, 45,
Oakland, California:

Neither a passenger in the strict commercial sense, nor a member of the Clipper's crew, Kenneth Kennedy was traffic manager for PAA's Pacific Division. He was flying west to the Orient on company business, possibly to reschedule the connections with CNAC, whose routes had been disrupted as the result of recent Japanese military advances. He surely didn't subscribe to Watson Choy's expressed view of a trip to China as being "fun." In his letter to "My Precious Wife," typed on a portable typewriter at 8000 feet above the Hawaiian Islands, Kennedy told her that he had "been busy every minute up to now, with not too much sleep. However, I will be able to take it fairly easy until I get to Manila." No typist, as he made clear in the first paragraph, the letter was amusing, in keeping with his innate sense of humor, but the humor seems forced and artificial and cannot entirely mask the dark concerns which seem to have haunted his mind.

Formerly employed by United Airlines, Kennedy was a family man who would be survived by his three children, Kenneth Jr., Marjory and Lynette, and his wife, Marjory. He would be mourned by his niece, Roberta, who appears to have thought the world of him, and by another woman, who loved him, as well, and who had dedicated a book of her poetry to him. Kennedy had drunk quite deeply from Life's cup, and, perhaps, he sensed that, ahead of him, were only dregs.

Although there was no evidence that the Clipper had actually been attacked, his wife, Marjory, was reportedly told (reliably, as she thought) that *Hawaii Clipper* had been shot down. But this could have been merely a logical assumption, based upon the June intelligence fears of an impending Japanese action, and upon the later facts—that the Clipper had, undeniably, been lost and that the *Kweilin*, had, indeed, been shot down.

Earl B. McKinley, MD, Colonel USAR, 43,
Washington, DC:

Married, and the father of two children, Dr. McKinley was the Dean of the College of Medicine at George Washington University, and had been a Professor of Bacteriology at the college since 1931. Widely recognized in medical circles, throughout the world, as an expert on leprosy, Dr. McKinley had studied this terrible, incurable disease in the Philippines, where, in addition to being field director for the Rockefeller Foundation, he had served as a member of the Governor

General's Advisory Committee on Leprosy and had lectured at the University of the Philippines. A graduate of the University of Michigan, Dr. McKinley had also studied at the Pasteur Institute of the University of Brussels, and had served as the Director of the School of Tropical Medicine in Puerto Rico. Aboard *Hawaii Clipper*, he described himself to K. A. Kennedy as a "specialist on tropical diseases." During World War I, he had held a reserve commission as a Captain in the Army Medical Corps, but had actually served as an Army intelligence officer.

In 1935, Dr. McKinley had published a study, *The Geography of Disease*, which outlined the general distribution of disease throughout the world. At this point in his career, he commenced his primary objective in medicine: that of determining the means of intercontinental transmission of disease.

In May, 1938, Dr. McKinley had isolated a new strain of the leprosy micro-organism in an American leper, and, in conjunction with his work with fellow passenger, Dr. Fred C. Meier, he was traveling to the Philippines to determine whether a study of skin tests on Filipino lepers might reveal the very same micro-organism which he had isolated in the United States. In addition, he and Dr. Meier were attempting to identify a vector for the transmission of leprosy, and other diseases, from Asia to North America.

Fred C. Meier, Ph.D., 45,
Chevy Chase, Maryland:

A botanist and, recently, a plant pathologist, from Framingham, Massachusetts, where his parents still resided, Dr. Meier had worked for the Department of Agriculture since his graduation from Harvard in 1915. From 1930 to 1934, he had been the principal pathologist with the U.S. Bureau of Plant Industry and had served, since that time, as a senior scientist in the extension service. Although Dr. Meier's field was botany, he shared a common interest with passenger, Dr. Earl McKinley, in determining how micro-organisms, including plant spores and disease bacteria, had found their way across vast ocean expanses. Neither of these men believed that commercial shipping played the major role, as science had long believed. McKinley and Meier looked—to the *winds*.

The Committee on Aero-biology of the U.S. National Research Council had been created to study the possible transmission of disease germs via the upper air currents, and Dr. Meier's work, in determining similar vectors for plant pollen and spores, had led to a joint enterprise with Dr. McKinley, sponsored by the NRC. Several years before, Dr. Meier had worked with Col. Charles Lindbergh to develop what they had called the *air-hook*. This was a sterile metal container with a retractable metal cover and a gelatin-coated glass slide, which, when attached to a long pole and extended into the air stream from an aircraft in flight, permitted the collection of airborne dust, pollen, spores and disease germs at attainable altitudes. It had proved to be more successful than the artificial heart that Lindbergh had developed, and one of these *air-hooks* is still retained by Purdue University, in the Amelia Earhart collection.

In 1937, Dr. Meier arranged with Amelia Earhart to have Fred Noonan use the air-hook to collect samples during her flight around the world, and, indeed, Meier was given an honorable mention in *Last Flight*, George Putnam's posthumous publication of Earhart's book. Presumably, the samples had been sent back to Washington following the Atlantic leg of her flight, because Meier had been able to demonstrate that at least forty micro-organisms, plant and disease, were being carried across the Atlantic on the upper air currents. With Earhart's loss in July, 1937, sampling had stopped, and Fred Meier had been unable to provide a similar study for the Pacific Ocean. However, with the National Research Council involved, Dr.

Meier, in July, 1938, was collecting his own samples aboard *Hawaii Clipper*, with the aid and collaboration of Dr. McKinley.

An International News Service report of July 29, 1938, noted that Meier and McKinley "did not reveal the specific nature of their current studies," but then went on to describe, in detail, both the air-hook and the gathering of data. Concluding with a clear reference to "the secret objectives of the scientific expedition," the INS news report indicated, quite openly, that there was more to this study than could be published.

[A related study was made in 1963 at the U.S. Coast Guard Loran Station at Johnston Island. Bob Sundell and other Smithsonian ornithologists were sent out to the station to determine the flight patterns of sea birds, which nested beneath the transmitting tower. Their cover story was that the Army was concerned that the Red Chinese might send disease-infected sea birds across the Pacific. Considering that a new island was being dredged up on the atoll at that time, and that the ornithologists were sending data to the Commanding General, Deseret Test Center (for biological warfare), it was apparent to a few of the Coast Guard crew that the new island was to be a *biological*, as well as a chemical, warfare center: the U.S. Army had been concerned that *it*, not the Chinese, might infect the far-ranging sea birds, and, sadly, this ultimately occurred, with tragic results, at Christmas Island, in the 1970's.

Years before, in May, 1945, Americans had learned of a Japanese air assault on the U.S., which had begun in November, 1944. Near Blye, Oregon, a minister had organized a picnic, and, as picnickers examined the large rice-paper balloon, which they had found, its payload had exploded, killing the minister's wife and five children. The press could no longer remain silent, but, while condemning the Japanese weapon as deplorable, made little of the balloon threat, perhaps because it seemed a last resort of a defeated nation, or because so few balloons made it to the U.S. According to the September, 1988, issue of *World War II Magazine*, this was due to the Japanese failure to prevent altitude-control batteries from freezing, up in the Jet Stream, a problem they would have solved by 1945.

Few Americans are aware, even today, that, while the U.S. focused on *atomic* weapons and the Nazis focused on *chemical* weapons, Japan concentrated on *biological* weapons of mass destruction—germ warfare—which would later become the nucleus of our own bacterial warfare arsenal, following Japan's defeat in 1945. The U.S. Government took the balloon threat seriously: a combat-ready fighter squadron was retained on the West Coast, solely to shoot these balloons down, and there was great concern for the effects of germ warfare on the U.S. population.

And for those who may wonder why we risked everything, in 1945, on two *prototype* atomic bombs, or why, from 1944 onward, the draft age was raised, basic training was shortened and raw recruits were shipped out of the U.S. at an astonishing rate, the answer should be clear: more balloons, plague—and millions of deaths—were expected in the late fall of 1945, when the 200 mph Jet Stream moved south out of Canada, and the flight, from Tokyo, to as far as Maine, might have taken as little as seventy-two hours. The consignment of disease to the winds, or even into the Jet Stream, is *still* something to think about, even as Meier and McKinley thought about it, aboard *Hawaii Clipper*, in July, 1938, or, as Amelia Earhart may have thought about it, somewhere, in July, 1937.]

Edward E. Wyman, Lt. Cdr., USNR, 45, Bronxville, New York:

Traveling on business, Ted Wyman was linked to events in the past and the future, which would finally reveal the tragic fate, in his own time, of PAA's *Hawaii Clipper*.

2. PASSENGER EDWARD E. WYMAN: The Wyman / Sutherland Connection

Hope Wyman (now Mrs. W. P. Streeter) was nineteen when her father, Edward Wyman, left home on a long business trip which was scheduled to transport him, quite literally, around the world. By all expectations, he should have been home in time for Hope's return to Pine Manor College, and to see his fifteen year old son, Corydon, off to prep school in Tucson. But, west of Guam, Ted Wyman suddenly found himself on a one-way flight to oblivion.

Edward Earle Wyman, "Ted" to his friends and associates—but never "Ed"—was forty-five years old and in the prime of both his life and his new career when he left his Bronxville, New York, home in late July of 1938. In May, after ten years as assistant to Pan American Airways' president, Juan T. Trippe, and facing a ceiling on the salary for his position, he had resigned from PAA. On June 3, he was named the Vice President of Export Sales for the Buffalo, New York, aviation firm, Curtiss-Wright Corporation, whose *Hawk 75* pursuit planes were being flown in China, in unequal combat against Japanese fighters and bombers. In late July, after barely a month and a half of meetings and company training, Ted Wyman found himself booked on a sales trip, which, most assuredly, involved business with China.

In fact, although Wyman's itinerary noted no more than a single day in any city from San Francisco to London, he had scheduled over a week in Hong Kong, or, more likely, in China, itself. His wife, Edna, may have wondered whether he planned to go into war-torn China, but Wyman seems not to have discussed any of his travel plans with her, especially those regarding the nine-day stopover in Hong Kong. But he did leave her a copy of his itinerary, which listed his Hong Kong address only as "c/o P.A.A."

Edward Earle ("Ted") Wyman, 1937

Snapshot by Edna (Wake) Wyman; photograph courtesy of Mrs. W. P. Streeter (Hope Wyman).

Although no longer affiliated with PAA, his use of the airline as an address was a typical practice of passengers in transit. Yet, PAA did not serve any of Wyman's further stops westward. These were served by Britain's Imperial Airways, and PAA could offer only extended air service into China, through its discreetly owned subsidiary, China National Aviation Corporation. One thing is certain: had Wyman booked nine days at a Hong Kong hotel, then he would surely have listed that as his address. But, as he appears not to have booked a hotel, then it may be assumed that he did not plan to remain in Hong Kong. In fact, Wyman's itinerary matched that of Major French, who, with Wah Sun Choy, had made no secret of an intended visit to China, during virtually the same period.

Moreover, it was also no secret that Curtiss-Wright had long been selling combat aircraft in China. Since before the outbreak of war, as *The New York Times* reported on July 31, 1938, China had purchased one hundred of the *Hawk 75*'s. These had been built in Buffalo of fabric and wood, which, at first, had appeared well suited to maintenance by untrained Chinese ground crews. But these fighters were never suited to combat in the hands of the poorly trained Chinese pilots and desperate American mercenaries, who had wasted as many as Japan had destroyed.

Yet, as this strangely appropriate *NYT* story also noted, something new was in the wind in the summer of 1938: China had switched to the all-metal *Hawk 75-A*, the 300 mph export version of the *P-36*, and intended to build them under license to Curtiss-Wright, which would sell its American-built engines to power the advanced new fighters. And, as Hope Streeter recalls, it was engines that were on her father's agenda in July, 1938.

The Chinese-assembled *Hawk 75-A*'s were to be equipped with armor plate behind the pilot's seat—the first in the world to be so armored. No longer would China have to depend solely on a pilot's luck for survival against the Japanese fighter-escorts These *Hawk 75-A*'s were fighters, as well, capable of prevailing against any Japanese aircraft then in production. And costly new fighters meant discipline, training and organization, as well as a new order of pilots, worth the weight of armor and skillful enough to fly with it aboard. For Japanese aviators, who had fared so well against China, the story could well have confirmed their fears of an impending loss of air superiority. And Ted Wyman was to be a party to that loss.

But, all career changes and foreign intrigue aside, if circling the globe by air would still be a fine trip, even today, it would have been a glorious adventure in 1938, and Ted Wyman was long overdue for an adventure.

Born in Evanston, Illinois, in 1893, Wyman had moved east with his family when he was nine years old. In 1908, about a year after his father's untimely death, he secured his mother's permission to strike out on his own, and, at the age of only fifteen, he had headed west. His travels took him from the Mississippi River to the Pacific and from the Mexican border to Canada. He worked cattle for a while and panned for gold, and, after working for a western railroad, Ted Wyman settled, for a time, on a 160-acre homestead near Clinton, British Columbia.

In 1913, at the age of twenty, he returned to the United States and moved east, settling, finally, in Waterbury, Connecticut, where he took a job at the American Brass Company. He caught up on his suspended education at the Roxbury School and eventually enrolled at the Sheffield Scientific School of Yale University, but it had been in Waterbury that he had first met Miss Edna Benton Wake. Her comfortable family background was in stark contrast to Wyman's rough and ready early years, but the two became engaged in 1914, and, by 1916, their circle of friends included a number of other students at Yale.

T. E. WYMAN
TENTATIVE ITINERARY

	July
San Fransisco-Honolulu	23rd
Honolulu-Midway	25th
Midway-Wake	26/27 (Date Line)
Wake-Guam	28th
Guam-Manila	29th
Manila-Hongkong	30th

Hongkong (c/o P.A.A.) July 30th-Aug. 7th

	August
Hongkong-Hanoi, Indo China	8th
Hanoi-Bangkok, Siam	9th
Bangkok-Rangoon, Burma	10th
Rangoon-Jodhpur, India	11th
Jodhpur-Basra, Iraq	12th
Basra-Athens, Greece	13th
Athens-Amsterdam	14th

Paris or London day or so later.

Ted Wyman – Sales Trip Itinerary, 1938

Itinerary given to his wife, Edna, by Ted Wyman, prior to his July departure from New York; copy courtesy of Mrs. W. P. Streeter (Hope Wyman).

Following his graduation from Yale in 1918, Ted and Edna were married, and, in 1919, after serving for a short time as a U.S. naval aviator, during World War I, Wyman went to work for the Mack Truck Company, becoming the National Sales Manager in the mid-1920's. Some three years later, in 1928, during the first full year of Pan American Airways' Caribbean operations, Ted Wyman joined PAA as an assistant to Juan Trippe.

Robert Daley, citing one of Wyman's own memos, casually refers to "the young men who served as assistants to the president," but Wyman was actually Trippe's senior, by six years, although they had actually been contemporaries, of sorts, at Yale University and, during World War I, as naval aviators.

Wyman and Trippe had both been at Yale in 1917, but Trippe was just beginning his first year and Wyman was about to be graduated. Trippe left for the U.S. Marine Corps in November of 1917, but later transferred to the Navy, to become an aviator. Wyman entered the Navy after spring graduation, in 1918, so it is likely that they first met at M.I.T., where both men attended aviation ground school. After flight training, they went separate ways until 1928, but Wyman remained active in the Naval Reserves, and, in 1938, held the rank of Lt. Commander.

Hardly an "office boy," as Daley so casually implies, Wyman was an educated, polished and successful businessman. He was also an adventurous and charming extrovert, who, almost from the very beginning of PAA's own adventures, provided a positive face for Trippe's reticent, secretive executive style.

A sportsman who loved flying, hunting and the Scottish games of curling and golf (with one term as the president of the New York State Golf Association), Ted Wyman also maintained ties to several engineering and aviation societies and retained an association with Yale through his alumni activities.

Despite an active personal life and extensive travel both in the Caribbean, and in Alaska, where he had supervised PAA operations, Wyman was a devoted family man, whose loss is all the more tragic, in that his fate has remained unknown for so long.

Because no trace of *Hawaii Clipper* was actually found—and because the ASB had left the case officially open—life insurance benefits would not automatically begin to replace incomes lost to family members. Earhart and Noonan, lost in July of 1937, were not declared legally dead until late in 1938, despite the massive naval air and sea search that was conducted, and it is unclear how long the Clipper's passengers' families waited, before receiving insurance benefits. [Of the crews' families, Eleanor Jewett, the widow of John Jewett, recently reported having received $50,000 from PAA.]

According to her daughter, Hope Streeter, Edna Wyman long believed that her husband would be found alive on some Pacific island and that, eventually, he would return to his family. But Life is seldom influenced by expectations, and Edna was soon compelled to give up her Bronxville home and to move to an apartment. Her daughter, Hope, could not afford to return to Pine Manor College, but remained with her mother and enrolled in a local secretarial school, obtaining the skills she required for gainful employment. However, the alumni of Wyman's class at Yale ensured that his son, Corydon, would complete his prep school education, before he, as his father before him, confronted a world war, with its risks and uncertainties.

After WWII, amidst the flood of revelations arising out of the long wartime silence, Hope was informed by her mother that word had come to her from "a friend, General Sutherland," that the engines of the *Hawaii Clipper* had been recovered—in Japan! It was, if true, a kind of death certificate, and, perhaps, from this time on, Edna Wyman no

July 14, 1980

Mr. William T. Seawell
Chairman & Chief Executive Officer
Pan American World Airways, Inc.,
100 East 42nd Street
New York, New York 10017

Dear Mr. Seawell:

I am writing this letter to urge Pan American to fully investigate the possibility that the wreckage found on Truk might be the missing Hawaii Clipper.

Perhaps you are too young to remember the accident but those of us involved have not forgotten. My father, Edward Wyman, at one time Assistant to the President of PAA, was aboard that clipper. You cannot imagine what it was like for my mother and the rest of his family not to know what really happened.

In "Search for Amelia Earhart" Fred Goerner mentions the wreckage of an unknown plane on an island in the South Pacific. I remember after World War II my mother heard from a friend, General Sutherland, U.S. Air Force, I believe, who also had heard about an unknown plane being found. At that time my mother talked to Mr. Robert Lord, then assistant to Mr. Trippe but nothing was ever done. Has Pan American ever done any follow up?

The recent book, "China Clipper" by Ronald Jackson makes a great deal of sense even though the ending is supposition. But is it really? Mr. Jackson hints of a Pan Am and Washington cover-up because the country was on the brink of war. Could this be a possibility? Does Pan Am know more than it ever said?

It occurs to me there must be former or present officials in Japan who would know if this hijacking took place and where the plane was taken and what happened to the people aboard. It is forty-two years since the disappearance -- surely this information could now be divulged.

There are three relatives (not including grandchildren) of Edward E. Wyman sill alive and I am sure our feelings are shared by any surviving relatives of the passengers and crew. We agree with Mr. Gervais that "it is the responsibility of Pan American to go to Truk; conduct a local investigation to determine if the fifteen people are there", and report their findings.

I would appreciate hearing from you.

Sincerely,

Mrs. W.P. Streeter

cc: Mr. James Arey
    Dir. Corporate Public Relations

Appeal for an Inquiry – Hope Streeter

In this appeal to Pan Am, the third paragraph is of some interest, with regard to Sutherland; copy courtesy of Mrs. W.P. Streeter (Hope Wyman).

longer waited for her husband to return. It was, if true, the kind of tragic news which no military officer would reveal casually, or without proper cause. And, considering that only Juan Trippe had received such post-war news of *Hawaii Clipper* (as Trippe, indeed, had received from Admiral John Towers, in 1946), it was, if true, the kind of classified information which only an extremely well-placed officer would have risked revealing.

There was such a "General Sutherland," of course, but Hope, along with many other Americans, had never heard of him, and, as late as 1980, she thought that he had served as an Air Force officer. It had been, perhaps, both a blessing and a curse to his career that Lt. General Richard Kerens Sutherland had served, throughout the Pacific War, as chief of staff to General Douglas MacArthur.

Having succeeded Eisenhower as chief of staff, Dick Sutherland had seen MacArthur through it all, from Corregidor to Australia, then the struggle to regain the Philippines, and finally, on to the occupation of Japan, in 1945. But there had never been light enough to illuminate anyone but MacArthur, alone. His staff, although as fine as any in either of the two wars, had served within the shadows of MacArthur's affected radiance.

Sutherland had been a Lt. Colonel in 1941, but, knowing that war was imminent and that the Philippines had been written off by the U.S., MacArthur had promoted many of his officers, creating an appropriate staff to endure what he expected would be a long and lonely Philippine command. In July, as the U.S. cut off Japan's naval oil supplies, Lt. Colonel Sutherland was promoted direct to Brigadier General, and on December 19, with the fate of the Philippines sealed, he was promoted to the rank of Major General. In 1944, in preparation for the invasion of the Philippines, Sutherland was promoted once again, to his final rank of Lieutenant General, which, for any graduate of West Point, would have been admirable, indeed. But it had been a remarkable achievement for Richard Kerens Sutherland, a graduate of Yale University—class of 1916.

It must have seemed quite strange to Edna Wyman's daughter, Hope, that only a lone civilian air tragedy widow, and, apparently, no one else, would have been entrusted with what appeared to be secret information. And not knowing who, or how highly placed, this "General Sutherland" might be, or anything else about him, for that matter, it would have been reasonable for her to have questioned the source of the report, as well as the lack of detail in her mother's account of it. But 1946 was not a year for looking back.

In 1964, the very year that Joe Gervais may be said to have "found" Ted Wyman on a Pacific island, Edna Wyman passed away. Hope Streeter came into possession of her mother's personal effects, including a photo album from happier days, early in 1916. In 1996, after speaking with the author, Hope examined some fading photographs that her mother had taken in 1916, at a house party at Fort Ticonderoga, on Lake Champlain. She found several in which appeared a young man, most certainly a friend, identified as "Richard Sutherland." At that time, Hope began to appreciate, more fully, perhaps, the validity of her mother's report, that traces of the Clipper had been recovered in Japan.

But now, knowing the university that Dick Sutherland had attended, the friendship that he had shared with Ted and Edna, the career that he had chosen to pursue, and just how highly placed he had been in Japan, in 1945, she can be assured, as can the reader, that her mother had been properly and accurately advised of the fate of *Hawaii Clipper*, and of Ted Wyman, by Lt. General Richard Kerens Sutherland, a man who had been, not only in a position to know all the facts, but secure enough in that position, and *loyal* enough, to risk revealing some of them to an old friend.

"Richard Sutherland," 1916

Snapped by Edna Benton Wake, during a house party near Lake Champlaign; Sutherland's name is shown, in quotation marks, as it was inscribed on the reverse of the original; from a copy print, courtesy of Mrs. W. P. Streeter (Hope Wyman).

Richard K. Sutherland, 1916

Photo and text from the *Yale Classbook* of 1916, reproduced by special permission of the Yale University Archives; Manuscripts and Archives; Yale University Library; New Haven CT; with the aid of Christine Weideman and Judith Schiff.

"R. Sutherland" (left), 1916

Snapped by Edna Benton Wake, during a house party near Lake Champlaign. Sutherland's name is shown, in quotation marks, as it was inscribed on the reverse of the original; from a copy print, courtesy of Mrs. W. P. Streeter (Hope Wyman).

Lt. Gen. Richard Sutherland (left), 1945

Sutherland witnessing as General Umezu signs the official surrender document for the Imperial General Staff, on U.S.S. *Missouri*, in Tokyo Bay, September 2, 1945; Signal Corps photograph; copy provided by Jack Novak, Col., USAF (ret.).

Venice, Florida 34285
August 8, 1994

TO WHOM IT MAY CONCERN:

I, Hope Wyman Streeter, hereby grant Charles N. Hill (or the bearer thereof) permission to search for, and to recover, if possible, the remains of my late father, Edward Earle Wyman, for return to the United States of America.

Hope Wyman Streeter

Sworn to before me this 8th day of August, 1994.

Deanna J. Horn

DEANNA J. HORN
Notary Public, State of Florida
My comm. expires Dec. 12, 1994
No. CC072872
Bonded Thru Western Surety Company

Recovery Authorization – Hope Streeter

Notarized letter, written for the author in 1994 by Hope (Wyman) Streeter, authorizing the search, recovery and return of her father's remains.

3. THE FLIGHT CREW

Howard L. Cox, 35, Engineer Officer
Certified Aircraft and Engine Mechanic
944 trans-Pacific hours

Cox had joined PAA in 1930, three years after Caribbean operations began. He served in Central America until 1937, and his longevity as a flight engineer was especially noteworthy, considering the standards that had been established by the Chief Engineer, Andre Priester. Along with the First Officer, Mark Walker, Cox had recently been in Seattle, assisting in the development of the new Boeing 314, and, like Walker, Cox had been sent back to regular flight duty after the new flying boat had exhibited disastrous characteristics during its maiden flight. His transfer to the new Atlantic run having been postponed, Cox was earning his pay back in the Pacific and, on Trip 229, he was training Tom Tatum, formerly a mechanic at PAA's Pacific island bases, but now, one of PAA's newest Assistant Engineer Officers.

George M. Davis, 29, Second Officer
Certified Transport Pilot
1650 flight hours/1080 trans-Pacific hours

An Ensign in the U.S. Navy Reserves, Davis hailed from South Lee, Massachusetts and had graduated from Massachusetts State College in 1931. After learning to fly in the Navy, Davis, as well as a good many others, including the First Officer, Mark Walker, had been released from active duty during the Depression cutbacks of the mid-1930's. He had joined PAA, in Miami, in 1935, but, in 1936, he transferred out to the Pacific and settled, with his wife, in Santa Barbara. As Second Officer, he also served as Navigator and was responsible for aircraft inspection, immediately after take-off. His wife, nearly nine months pregnant, was hospitalized soon after the Clipper's disappearance.

John W. Jewett and family, 1933

Jewett, with wife, Eleanor, and daughter, Anne, in Miami (a son, John, was born in 1935); photo courtesy of Anne's daughter, Jan O'Sullivan.

John W. Jewett, 27, Fourth Officer
Certified Transport Pilot
2000 flight hours/428 trans-Pacific hours

John Jewett, Lt. (j.g.), USNR, from Milton, Massachusetts, was a graduate of M.I.T., where he had been a record-breaking track star. He received flight training in the Navy, and, after working as a Douglas engineer, he joined PAA in 1936 and flew the Caribbean routes, out of San Juan. In 1937, he was transferred to the Pacific, and, although a more experienced pilot than any of *Hawaii Clipper*'s officers, except Terletsky, his lack of Pacific flight time had relegated him to a training berth, as Fourth Officer.

William McCarty, 33, Flight Radio Officer
Certified Radio Operator (First Class)
1352 trans-Pacific hours

It was reported in the press that McCarty had been Chief Radio Operator aboard the S.S. *Lurline*, a passenger liner well known in Pacific service, between San Francisco and Hawaii. As an old hand at sea, McCarty may be assumed to have been, not only a radio operator of invaluable experience, but a man who was not unfamiliar with the darker side of life in the Pacific basin. It is not known whether anyone survived him, other than his mother, Mrs. Elizabeth McCarty Coop.

Ivan Parker, 40, Flight Steward
1200 trans-Pacific hours

A veteran of some 26 Pacific trips, Parker had been living on borrowed time since January, when, due to schedule conflicts, he had missed joining the ill-fated mail flight of the *Samoan Clipper*, which was lost, with all aboard, near Pago Pago, American Samoa. According to Jackson, Parker had given up on his dream of becoming a Flight Radio Officer and planned to accept a promotion as manager of PAA's hotel on Guam. Ruth, his wife, refused to trade spousal absence for isolation and declined to relocate to Guam, so, shortly before *Hawaii Clipper* departed, Ivan Parker had filed for divorce. Jackson relates that Ruth changed her mind just after his departure—but it no longer mattered.

Jose M. Sauceda, 35, Third Officer
Certified Transport Pilot
1900 flying hours/570 trans-Pacific hours

Originally from Texas, Sauceda had been a PAA employee longer than had anyone else aboard, other than Leo Terletsky and Ted Wyman. He had flown for PAA since 1929 and, in 1931, he had been co-pilot on PAA's charter flight into Brazil's "lost world," the Matto Grosso. He had remained in Brazil until late in 1937, when he transferred to the Pacific Division. Walking to the dock on the day of *Hawaii Clipper*'s departure from Alameda, Sauceda was overtaken by Mark Walker, driving his new Ford. Declining to squeeze in behind Walker's younger sister, Mary Ann, Sauceda, with casual flair, rode in, standing on the running board.

Thomas B. Tatum, 29, Asst. Eng. Officer
Certified Aircraft and Engine Mechanic
35 trans-Pacific hours

Although he had logged only a few hours of trans-Pacific flight service (all on Trip 229), Tatum was a Division pioneer in the truest sense, having served, from the beginning of PAA's Pacific operations, at the airline's lonely outposts out on Wake and Midway, as well as at the more populated station at Guam. Tatum had performed regular ground maintenance on the Clippers for three years and had only recently been promoted to flight duty. He had joined Trip 229 at Pearl Harbor and had begun his in-flight training under the Engineer Officer, Howard Cox.

Leo Terletsky, 43, Captain
Certified Transport Pilot
9200 flying hours/1626 trans-Pacific hours

Born in Samara, Russia, in 1895, Terletsky left Russia after the Bolshevik victory and eventually immigrated to America in 1919, graduating from Columbia University and becoming a naturalized U.S. citizen in 1925. Terletsky had begun flight training in 1920, at Curtiss-Wright Field on Long Island, and, from 1922 to 1924, he had barn-stormed in the east. In 1926, he worked as a pilot for WACO, a maker of aerobatic biplanes, and, after a short stint as Chief Pilot for Barrett Airways in 1927, he came to PAA in 1928. His first assignments with PAA were in the Caribbean and in South America, but he was temporarily assigned to New York in 1932, as PAA began exploring new routes. During his New York assignment, Terletsky married Miss Saretta Bowman, a history teacher and the dean of non-resident students, at Sarah Lawrence College. For a few months, they lived in Bronxville, New York, but later moved to Miami and, later still, to Yucatan. In 1936, when Terletsky transferred to the Pacific Division, they settled in Palo Alto.

On July 29, 1938, when two Japanese naval officers hi-jacked *Hawaii Clipper*, it wasn't the first time Terletsky had faced an armed and unwelcome passenger. According to Daley, in August of 1933, Terletsky had been in Cuba when a revolution broke out. Secret Police Chief Ferrara and another Cuban boarded the Clipper, intending to escape to Miami, but PAA Station Manager Kauffman, fearing retribution from the revolutionaries, wanted no part of them and demanded that they disembark. Ferrara then pulled a gun and refused to depart, so Kauffman had the Sikorsky *S-42*B towed to the mooring buoy, out of the way, where Terletsky could start the cold engines.

As Terletsky attempted to start the balky engines, a mob arrived at the station dock, whereupon Kauffman waved a red flag at Terletsky, as a signal to take off. With only one cold engine running at idle, Terletsky, who was highly skilled in "sailing" a flying boat (i.e., maneuvering it in the wind on the surface), began working the Clipper out to the take-off area. At the station dock, the revolutionary mob thought Kauffman's red flag was a signal to *them* and turned their attention to the Clipper. As the other engines were started in turn, Terletsky continued to "sail" the Clipper at low throttle, even after the mob began firing at the flying boat. Only after nine bullets had pierced the fuselage, one of them just six inches from Terletsky's head, did the four engines suddenly rise to full throttle, pulling the plane out of range. This, for the most part, was the story that Daley related, one that gave the impression that Terletsky was unflappable in a crisis. What Daley did not relate, or else had never heard, were the details which had originated with the flight crew and which were then circulated throughout PAA. According to this version (which was told to the author in 1992 by a former PAA flight officer), as the mob began firing at the aircraft, Terletsky appeared to "freeze up," continuing to "sail" the Clipper with the engines just above idle. As the story went, he seemed oblivious, both to the gunfire and to the loud exhortations of his crew. According to this version, it fell to his co-pilot, finally, to run the engines up and to taxi the Clipper out of harm's way.

The second version seems consistent with a negativism that appears to have hounded Terletsky within PAA. The retired pilot also recalled, in 1992, that Terletsky had been known as the "Mad Russian," a demeaning nickname, but predictable at a time when Russians were stereotyped as high-strung, manic types. Others remembered him as a Captain who spent far more time with the passengers in the lounge, than with the flight officers on the bridge, but then, PAA was expanding in the late 1930's, and the need for new flight crews had made it necessary for pilots, such as John Jewett, on Trip 229, to spend as much time as possible at the flight controls on the bridge, building time and experience. Terletsky's official record seems to speak for itself, but there is a shadow which surrounded him at PAA and which may never be fully illuminated.

Mark Anderson Walker, 26, First Officer
Certified Transport Pilot
1900 flying hours/1575 trans-Pacific hours

A Texan from Wichita Falls, Walker was a Naval Reserve Ensign, whose *curriculum vitae* was much the same as that of Davis and Jewett. But only Walker has left behind, through his sister, Mary Ann, and his first cousin, Captain Robert Greenwood, a legacy of personal observations, which, after a long silence of over sixty years, has finally begun to shed light on the very real concerns and fears of PAA's Pacific Clipper crews.

Amelia Earhart

Shipmate: Referring to "The Search for Amelia Earhart" — Capt. William B. Short, Jr. USN (Ret.) — *Shipmate* November 1986:

This article presents a most interesting account of one aspect of the Earhart story, the search. It also brings to light a common situation where the participants in a naval operation may not be privy to intelligence information concerning their activities. Apparently Capt. Short and his shipmates were not aware of the true circumstances of Amelia Earhart's daring flight and the more likely position of her disappearance.

In the referred article, Captain Short's 5 July '37 letter vents emotion about this "publicity stunt" and its effect on public confidence.

In the Summer of 1938 my first cousin, Mark Walker, was visiting the family in Texas. He and I had a long discussion about his life in the Navy, flying off SARATOGA in the early thirties, his employment in aerial photography, experiences as a Pan American pilot in Sikorsky flying boats in Central and South America, his current life as a China Clipper first officer and of his test flights in the new Boeing YANKEE CLIPPER which he was to captain on the Atlantic route. His visit home was possible because the test seaplane had suffered sabotage, and the schedule of test flights had to be delayed.

Since Walker was convinced that Japanese agents were responsible for the damage to the Boeing Clipper, the subject was raised about possible sabotage by Japan to one of the Martin China Clippers and Mark talked about the Earhart disappearance. He was convinced that she had been forced down by the Japanese. And, his opinion was much more than guesswork.

Early in 1937 Mark had been assigned to work with Amelia Earhart and her navigator Noonan on their Pacific area phase. He at once urged his friend Earhart not to risk the emphasis that Pan American placed on flight safety by such a foolhardy "publicity stunt." He told her that her equipment was barely adequate.

Her reply was direct. She had not proposed the flight. Someone high in the government had personally asked her to undertake the mission. Her navigator was an accomplished aerial photographer. The flight was to be laid out with two routes. One was to be publicized. The other was to be directed over intelligence objectives in the islands controlled by Japan. Positions on the published route could be translated to positions on the actual route.

As a side note, Mark Walker described how he and his fellow Pan American pilots had discussed how easy it would be for a saboteur to sneak aboard a China Clipper and with no more than a pistol commandeer the flight and direct it to another destination. The clippers had all of the latest Navy instrumentation and communications equipment which he felt the Japanese wanted.

About a month after his visit home Mark substituted for an ailing pilot as first officer in PHILIPPINE CLIPPER in a flight leg from Guam to Manila. The Clipper disappeared at the nearest point to a Japanese controlled island. Their position was known because their radio transmission stopped abruptly after reporting fair weather and their precise location. Contact could not be reestablished although radio conditions were good. Walker's prophetical conjecture had apparently come true. The cargo on that particular flight was Chinese gold bullion and a few high Chinese officials, including, I believe, the defense minister.

In the years that followed several visits were made to Mark Walker's father by Naval Intelligence officers. However, his father would never reveal the purpose of these visits because he had agreed to secrecy.

As regards the Earhart search, the Navy obviously knew more about the flight than was communicated to the search force participants. Perhaps the *misguided* search was a planned public diversion to reinforce the image of the United States as a peaceful, non-spying nation. This attitude apparently still covers other unpublicized intelligence probes into the Mandated Islands that were conducted in the pre-war period.

R.B. Greenwood '43

The Greenwood Letter

Reproduced by permission, US Naval Academy Alumni Association: *Shipmate* (by David Church)

Recollection of conversations with Mark Walker, *Hawaii Clipper* First Officer, by his first cousin, Captain Robert Greenwood (clippings from the Jan.-Feb., 1987, edition of *Shipmate Magazine*).

4. FIRST OFFICER MARK WALKER: The Greenwood / Lee Reflections

A key point, with regard to the analysis of *Hawaii Clipper*'s last flight (for PAA), concerns the work done, presumably by the Navigator, Second Officer George Davis, to "get the word out" by means of surreptitious reconfigurations of the false position reports. Although a remarkably inspired effort, the limited time available certainly required, if not advance preparations, then, at the least, advance *expectations* of a hi-jacking.

Ronald Jackson's *China Clipper* states that, "in late June through early July, Acting Secretary of Commerce Colonel J. Monroe Johnson had heard rumors that one of the Clippers might soon be sabotaged." This came from records and informants whom Jackson trusted, yet he also added that, "the rumors died down," and concluded, perhaps without substantiation, that, "certainly those flying aboard the Clippers knew nothing about them." Of course, it is unknown as to whether or not the Clipper crews had direct knowledge of the warnings which circulated at Commerce, but, direct knowledge or not, PAA's flight crews were surely aware of the dangers posed by the Japanese. These crews, after all, worked at the "front lines" for PAA and had better cause than had the Commerce Secretary, to be sensitive to tensions in the Pacific, and this is more than an assumption:

There are two sources which indicate that one of *Hawaii Clipper's* crew, First Officer Mark Walker, had not only expressed deep concern, as to the danger of Japanese action against a Clipper, but had actually defined that risk in terms of a possible *hi-jacking* (although that particular term would not be applied to air piracy until the mid-1950's). Walker was not alone in foreseeing Japanese action against PAA and, as will be seen, had personally confronted the reality of Japanese aggression on one desperate occasion.

Mark Anderson ("Tex") Walker, 1938

Snapshot by Ralph Harvey; photo courtesy of *The Times Record News*, Wichita Falls, Texas, from front page story, "Last goodbye," by senior staff writer Lois Lueke, published April 16, 1992.

The process of gaining insight into Walker's concerns began, for the author, in 1987. In its January-February edition of that year, the U.S. Naval Academy Alumni Association's *Shipmate* published a letter from Captain Robert B. Greenwood, USN (Ret.), class of '43, which had been written in response to an earlier letter condemning Earhart's last flight as a publicity stunt. In addition to the author's notes, the excerpts contain details, embedded in brackets, provided by Captain Greenwood, in 1987. It should be noted that, since 1990, the retired U.S. naval officer has declined to discuss the Clipper hi-jacking.

An extensive discussion of the details of Walker's reported encounter with Earhart, which follows, has been provided because it

21

is especially unique. Many researchers have either indicated, or attempted to prove, that the last flight of Amelia Earhart was, in fact, a covert intelligence operation undertaken in the interest of American national security. Walker's story is one of the few, if not the only, account (albeit hearsay), in which she is alleged to have *admitted* to be preparing for an intelligence flight over the Japanese Mandates. Because the account is unique, in this respect, it is necessary (if Walker is not to be dismissed out of hand) that the Earhart spy-flight scenario, which he described to Greenwood, be seen as technically viable:

"In the summer of 1938 [three weeks before the *Hawaii Clipper*'s loss] my first cousin, Mark Walker [First Officer on the Clipper], was visiting the family in [Wichita Falls,] Texas. He and I [then fifteen years old] had a long discussion about his life in the Navy, flying off Saratoga in the early thirties, [his discharge, during the Depression, along with other Reserve Officers,] his employment in aerial photography, experiences as a Pan American pilot in Sikorsky flying boats in Central and South America, his current life as a China Clipper first officer and of his test flights in the new Boeing [model 314] Yankee Clipper which he was to captain on the Atlantic route. His visit home was possible because the test seaplane had suffered sabotage, and the schedule of test flights had to be delayed."

[Jackson makes no mention of any sabotage affecting a 314 boat at this time, although his research had centered, in large part, on FBI records of sabotage investigations (and had revealed the Commerce Department rumors). Daley, however, in *An American Saga*, notes a near-tragic initial test flight of the 314, in which the wing's high dihedral angle and the small single vertical stabilizer had prevented the aircraft from turning in flight, a design deficiency. This followed soon after the boat's launching, on June 3, 1938, during which it nearly capsized, also due to a design deficiency. Boeing may have remained silent, thus permitting suspicions of sabotage to explain the reworking of the flying boat, but, in any case, sabotage or not, Walker apparently had sabotage in mind.]

"Since Walker was convinced that Japanese agents were responsible for the damage to the Boeing Clipper, the subject was raised about possible sabotage by Japan to one of the Martin China Clippers and Mark talked about the Earhart disappearance. He was convinced that she had been forced down by the Japanese. And, his opinion was much more than guesswork.

"Early in 1937 Mark had been assigned to work with Amelia Earhart and her navigator Noonan on their Pacific area phase. He at once urged his friend Earhart not to risk the emphasis that Pan American placed on flight safety by such a foolhardy 'publicity stunt.' He told her that her equipment was barely adequate."

[He apparently told her in no uncertain terms, as Captain Greenwood informed me in 1987, and as the candid language of his clear recollections would appear to confirm. Walker was apparently displeased that PAA had risked its strong reputation for safety by involving itself with what still appears to have been a risky flight for a small land-based aircraft. He apparently feared that the growing public confidence in trans-Pacific flight might be shaken, if she were lost.]

"Her reply was direct. She had not proposed the flight. Someone high in the government had personally asked her to undertake the mission. Her navigator was an accomplished aerial photographer. The flight was to be laid out with two routes. One was to be publicized. The other was to be directed over intelligence objectives in the islands controlled by Japan. Positions on the published route could be translated to positions on the actual route."

[This was not only technically possible, but also consistent with anomalies in Earhart's flight from New Guinea. As to the technical possibilities, the published routes specified a ground speed of 150 mph, yet Earhart's own notes, written during the March 17, 1937, flight to Hawaii (and available to researchers at Purdue University), indicated a speed of, "180 mph Boy oh boy..." but, as she later noted, they had "...throttled down to 120 indicated airspeed so as not to arrive in darkness." Moreover, the text of *Last Flight*, largely ghost-written by publisher George P. Putnam, her husband, noted that "actually, we were going about as slowly as possible. We throttled back the engines and most of the way our craft was 'under wraps'."

"Under wraps" was a curious statement, to say the least. Amelia Earhart and her three companions, in setting a record for Oakland-to-Honolulu at 153 mph ground speed (near her publicized speed of 150 mph), had not only "throttled back" to keep the *Electra*'s capability "under wraps," but had set this record with the right propeller jammed in a fixed pitch-position, and the left propeller pitch-control operating stiffly. (As Air Corps records indicate, both prop spiders had been "lubricated" with a putty-like contaminant).

Later, during Earhart's second (and final) world flight attempt, the *New York Times* reported her speed, from San Juan, Puerto Rico down to Carripito, Venezuela, as being nearly 190 mph (true air speed, that is, for a ground speed, against headwinds well above 30 mph, of just over her desired 150 mph average). The *NYT* also cited a top air speed of *250 mph*, which makes it apparent that, whatever her ultimate plans may have been, Earhart could have appeared to make a flight from Lae, New Guinea to Howland Island, at 150 mph ground speed, but while actually detouring to Truk, in the Japanese Mandates, in the same time—but at higher air- and ground-speeds, which, for the most part, were, understandably, kept "under wraps."

What Earhart told Walker regarding this spy flight, while clearly serving to "put him in his place" for his criticism, was, technically, quite feasible. And, if Walker's comments were abrasive, as Captain Greenwood has indicated that they may have been, then her "direct" reply, while constituting a serious breach of security, can easily be seen as an understandable, if careless, rejoinder. The omission of a book credit for Walker would be consistent, as well, with several reports of Earhart's unforgiving temperament.

Most important, there is a "ring of truth" to the detail regarding a translation, or tie-in, of *reported* positions, to *actual* positions along a secret route. In 1985, the author found that Earhart had the speed and fuel to fly to Truk, en route to Howland, but *could not* include Mili Atoll and still reach Howland with the fuel and time available. The tactic served the hi-jackers of *Hawaii Clipper* far better than it served the Earhart spy-flight planners.]

"As a side note, Mark Walker described how he *and his fellow Pan American pilots* had discussed how easy it would be for a saboteur to sneak aboard a China Clipper and with no more than a pistol *commandeer the flight* and direct it to another destination [italics added by the author]. The clippers had all of the latest Navy instrumentation and communications equipment which he felt the Japanese wanted.

"About a month after his visit home Mark substituted for an ailing pilot as first officer in Philippine [actually *Hawaii*] Clipper in a flight leg from Guam to Manila. The Clipper disappeared at the nearest point to a Japanese controlled island [not flying *west!*].

"In the years that followed several visits were made to Mark Walker's father by Naval Intelligence officers. However, his father would never reveal the purpose of these visits because he had agreed to secrecy [Mark was never after discussed at home]."

Speaking with Captain Greenwood in 1987, I learned that he and Mark Walker had also discussed Greenwood's hope of becoming a Navy flier. Walker, a Navy Reserve Ensign and pilot, who had lost his active duty status, due to Depression cutbacks, advised his first cousin to graduate from the Naval Academy, as a first step toward a secure career in naval aviation—which is exactly what he did.

Captain Greenwood's letter and subsequent reflections on his conversation with Mark Walker, while providing valid speculation regarding Earhart's last flight, also confirms, not only that PAA flight officers were aware of the possibility of a hi-jack attempt, but that at least one of them believed that a Clipper hi-jacking might well be successful.

For men to function under the darkest clouds of destiny, there must always be hope, if not to prevent an act, then, at least, to mitigate its effect. Given the fear that a Clipper could actually be hi-jacked, it is apparent that the only real hope left to PAA flight crews was that, if hi-jacked, they might, somehow, "get the word out," as to what had happened. Among PAA flight personnel in the Pacific Division, some course of preparing for a hi-jacking surely evolved from an assessment of their assets and liabilities. But the hope of "getting the word out" was not, necessarily, the pointless action that it might seem to be:

It was a hope which carried with it a further hope of rescue: by the U.S. Navy, eager to draw the line on Japanese hostility in the Pacific; by embarrassed Japanese diplomats, eager to forestall war with the United States or by rational senior officers of the Imperial Navy, eager to continue sending their naval tankers to Long Beach—for oil to fuel the Japanese fleets. It was a hope, too, which could free them to face a deadly showdown with a hi-jacker and, perhaps, beyond that, free them of that crippling human terror: that they might simply vanish, their fates known only to God—and to their murderers.

A graduate of Stanford University, Walker had trained as a carrier pilot at Pensacola, and, after leaving the Navy, became an avid underwater, as well as aerial, photographer, before coming to Pan American Airways. Following his visit to Wichita Falls, Texas, in early July of 1938, Walker received word that another Flight Officer had fallen ill and that he would be required to fill in as First Officer aboard *Hawaii Clipper*, on Trip 229. Although he was scheduled to inaugurate the new Atlantic route, delays in development of the Boeing 314 boat, the *Yankee Clipper*, had made him available as a relief pilot. In any case, he needed a few extra flight hours to qualify for command of the new Clipper.

On his return to Oakland he had brought along his twenty-year old younger sister, Mary Ann, intending to show her around the Bay Area. Although disappointed by Mark's unexpected orders, which would leave her alone and on her own for three weeks, she was to have the use of his home and his new Ford—and the run of the Bay Area. Before he departed, however, he still wanted to spend some time with her, and so, she was privileged to observe PAA operations from the "inside," where she met several of the pilots, including Captain Leo Terletsky and Second Officer (and also Navigator) George Davis, whose wife was expecting their first child—due barely a week after Davis was scheduled to return from the Pacific.

On the day of departure, as Mark and Mary Ann approached PAA's terminal facility at Alameda, Walker stopped at the gate to pick up the Third Officer, Jose Sauceda, who was walking to the Clipper dock. As Mrs. Mary Ann Lee recalls, Sauceda accepted the lift, but declined, considerately, to squeeze into the small Ford, and rode in, standing on the running board. An hour or so later, the big flying boat pulled away from the dock, and soon afterward, on its take-off run, Mary Ann was thrilled to see Mark look out of the cockpit and wave a second goodbye to her.

But, as she also recalls, neither Mark, nor another of his fellow Flight Officers, were particularly sanguine about the flight. On the eve of departure, Mary Ann had spent the evening with her brother and another PAA pilot and his wife. Later, after a fine time on the town, the four had returned to Mark's home, where the men had become engrossed in PAA shop talk. Relaxing on the couch, Mary Ann soon became aware that the mood of the two Clipper pilots had grown somber, and, although she cannot recall all that was discussed, she remembers that they spoke at length, and in hushed tones, of Captain Ed Musick, and of his *Samoan Clipper*—blown to bits six months before, near Pago Pago.

A day or so after *Hawaii Clipper* was lost, Mary Ann Walker sat in an office at PAA's Alameda terminal and listened as an official explained that the Clipper had been lost in a typhoon. As he spoke, she raised her eyes to a Pacific map behind the man's desk. She saw the time zone clocks, the air route and all the weather patterns drawn upon it, but nowhere along the route could she find any sign of a typhoon. She couldn't understand why he would lie to her, but as she looked back at the official, intending, in her Texas fashion, to call him on his lie, he averted his eyes and lowered his head. Mary Ann felt that, for some reason, he had been ordered to lie to her, and, considering his concern, with regard to suppressing contact with the press, PAA may well have had something to hide.

During the next week, while living alone in her brother's home, *and under PAA orders not to speak to the press*, Mary Ann found herself dodging reporters at every turn, even having to climb onto the neighbors' balcony and to sneak out through their entrance. But, a few weeks after *Hawaii Clipper* was lost, when it was sadly apparent that Mark would not be returning, Mary Ann began to take stock of the loose ends in her brother's life. On his desk, amidst a growing accumulation of letters and telegrams, she found a note that Mark had written to her, just prior to his departure. It had to do with his new Ford and, in it, he reminded her to have the tires rotated—*some six weeks later*! Mark Walker had not expected to return from the Pacific.

Although he was concerned about a possible hi-jacking, Walker had a personal reason, as well, to be uneasy about the Japanese in the Pacific. As he told Mary Ann, he believed that he had been marked, for retribution, by the Japanese, for his role in protecting the facts, regarding the Japanese naval air strike which sank the U.S. gunboat *Panay*, on the Yangtze River in China, in December, 1937:

Japan had initially apologized for what they called an "unfortunate mistake," but radical Japanese officers, anxious to protect their careers, accused the *Panay* of having fired the first shots—a ridiculous charge, but one which had to be answered. Of course, after the *newsreels* of the attack were shown in theaters throughout the U.S., the lies became untenable. Japan agreed to pay, and finally paid, an indemnity for the *Panay*'s loss.

The air attack had been filmed by fast-acting newsreel cameramen aboard the *Panay*. The films (and photos) had been protected by the survivors, before being taken down-river to Shanghai and from there, aboard the cruiser U.S.S. *Augusta*, to Manila. With Americans eagerly awaiting all the facts, the U.S. Navy decided to send the films by Clipper, and the responsibility for guarding them fell to Mark Walker, a Navy Reserve Ensign. On his way to PAA's Cavite base, to board the *China Clipper*, he was accosted by several young Japanese, dressed in civilian clothing. They demanded that Walker hand over the cans of film, but he refused, of course, and, just as it appeared that things were about to become violent, a few bystanders began moving to Walker's aid. As the Japanese retreated, one turned and terminated the encounter with a chilling farewell, saying grimly, "We know who you are—*Mark Walker*!"

Munday, Texas 76371
May 30, 1995

TO WHOM IT MAY CONCERN:

I, Mary Ann Walker Lee, hereby grant Charles N. Hill, or the bearer thereof, permission to search for, and to recover, if possible, the remains of my late brother, Mark Anderson Walker, for the return to the United States of America.

Mary Ann Walker Lee

Sworn to before me this 2nd day of June, 1995.

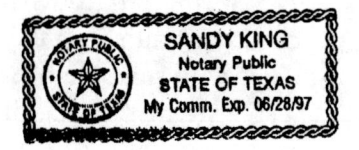

Sandy King, Notary Public.

SANDY KING
Notary Public
STATE OF TEXAS
My Comm. Exp. 06/28/97

Recovery Authorization – Mary Ann Lee

Notarized letter, written for the author in 1995 by Mary Ann (Walker) Lee, authorizing the search, recovery and return of her brother's remains.

5. THE HI-JACKERS

There is little doubt that, while the Pacific War was largely a result of political efforts by the Fleet Faction of the Imperial Japanese Navy—which had been planning for a naval surface war against the United States since 1911—it gained a great deal of momentum from the *ad hoc* operations of naval aviators.

These included the known air attacks on the U.S. gunboat *Panay* and the CNAC *DC-2*, the *Kweilin*, as well as the less well-known operations involving Earhart, the destruction of the *Samoan Clipper* and the hi-jacking of the *Hawaii Clipper*. Radical Japanese naval aviators operating independently (with some senior officer support) within the large Fleet Faction implemented all of these operations. They sought to alter the original naval plan, by which the U.S. Navy was to be drawn into a decisive surface battle in Japanese home waters and hoped to create a far more modern plan, in which the Imperial Japanese Navy would use air power to challenge the American Navy—*in American waters*. But to achieve these ends, they had to re-direct Japanese policy (then under army influence), which favored a joint Axis land war against the Soviet Union, with the Nazis acquiring the breadbasket of the western Soviet states and the Japanese acquiring the raw material treasure trove of Soviet Siberia.

What Japanese naval aviators sought was less a supporting role in a land war in Asia and more a glorious naval air war against the United States, whose spheres of influence, in the Pacific, intersected their own. What they wrought, in the end, was the destruction and the humiliation of Japan. In seeking war, for its own sake, they, *and those who aided them*, became, in a real sense—lost souls.

In the late 1930's, this ambitious quest for a "glorious" naval air war in the Pacific was led by the Imperial Japanese Navy's radical aviator, Lt. Commander Minoru Genda. Yet, in 1945, it could not have been lost upon this scion of an old Samurai family that, in his home prefecture of Hiroshima, there were far fewer Japanese to appreciate his glory.

As for his friend, Lt. Commander Mitsuo Fuchida, who was the first pilot over Pearl Harbor on December 7, 1941, and the last to depart, a post-war conversion to Christianity and a life of evangelism is not at all difficult to understand. These men, and those who flew with them, would have done well to consider the observation of General William Tecumseh Sherman, that, "boys think war is glory, but, boys, war is hell!"

Japanese naval aviators, by mid-1937, were highly motivated, but were still undervalued. At the Japanese Naval Staff College, Genda, according to the *Encyclopedia of Military Biography*, "was known as Madman Genda for his obsession with the primacy of naval aviation." Although he shared, with others, the dream of a naval victory over the U.S., his vision was focused on a victory for naval *aviation*, and his grand obsession was only intensified by the lack of interest in aviation, which then typified the Naval Staff College.

As Gordon Prange's group writes, in *God's Samurai*, "no first-class fliers taught at the Naval Staff College, although it had excellent instructors in gunnery, torpedo warfare, and submarines." It was inevitable that Genda and his associates, having had to become their own *sensei*, or masters, at the Staff College, soon perceived themselves as *ronin*, or, masterless Samurai. The legendary leader of the *Forty-Seven Ronin* had masked his quest for vengeance in feigned madness, and so, it may be that Genda took no offense at being known as "Madman" Genda.

The resentment of these naval aviators was not founded upon shortsighted conventional thinking, alone: As Prange defines it, "The difference in strategy between Japan's army and navy was surprisingly fundamental. The navy thought in terms of...battle with the United States; the army pointed toward a major land conflict with the Soviet Union...Broadly speaking, therefore, the struggle in China was an army show. No major naval battles thundered along the China coast, and Japan's surface ships had little difficulty dominating what action there was." Prange does add that, "the naval air arm, however, found many opportunities for service from land bases or from carriers."

But there was little *glory* for the Navy in providing support for the Army's faltering conquest of China, nor would there be any at all in standing by, while the army tried to cut Siberia away from the Soviet Union. Given the Army's projected failure against Siberia, however, Japan's naval aviators needed only an air superiority fighter, a viable plan for an air assault on the U.S. Navy, the means to wage a long-range air war at sea and an icon of American spiritual decadence (to sanctify their crusade), and they might force the re-direction of Japanese policy and provoke the U.S. into war through an escalating series of hostile acts, culminating, if necessary, in an attack on the U.S., itself, or on its territories.

In 1937, they apparently acquired the very tools that they needed, because it was late in that year that they took the first step to force an air war with the U.S. The step was a naval air attack on the U.S. gunboat *Panay:*

Tracing Fuchida's path to war, Prange notes that, "Fuchida became involved in the China Incident [*Japan's* euphemism for 'the war in China'] after Japanese fliers pounced on the U.S. gunboat *Panay* in the Yangtze River near Shanghai [actually *at* Nanking, *during the 'rape' of that city*, as Prange's 'editors' well knew] in the early afternoon of 12 December 1937." Fuchida was then attached to the Naval Staff College, but was ordered to report to the Thirteenth Naval Air Corps at Nanking, as Chief Flight Officer. Prange recounts Fuchida's briefing for the mission, which was issued by Rear Admiral Zenshiro Hoshina. As Fuchida recalled, this briefing clarified the truly renegade nature of the *Panay* attack, which had shaken Japan as much as it had America:

Fuchida was briefed thus: "'An event like this Panay incident is seriously detrimental to our war effort,' he [Hoshina] emphasized. 'It appears that two young lieutenants named Murata and Okumiya were...the ringleaders. You will be responsible for controlling these undisciplined officers by all means. They must not get out of hand again and upset our critical international relations.'" However, this briefing should not, necessarily, be construed as Hoshina's rejection of the *spirit* of the renegade attack, only its *timing*. In commenting on the orders, Prange adds that, "whatever his sentiments toward the two fliers at the time, Fuchida soon counted them among his friends."

Shigeharu Murata (who perished in the Battle of Santa Cruz in October of 1942) became an expert in torpedo bombing tactics and led the torpedo bombers in the first wave at Pearl Harbor. It was he who had developed the wooden fins, which permitted torpedoes to be used in the shallow waters of Pearl Harbor. The other "ringleader" in the *Panay* attack was Masatake Okumiya, who survived the Pacific War and, with Fuchida, co-authored a book on the battle of Midway, published in 1955 by the United States Naval Institute Press (which has refused to publish this work). In 1937, the two airmen, although renegades, to be sure, were, by no means, common brigands, but, like Fuchida, were dedicated professional naval aviators. And, as might well be expected, Fuchida's mission in Nanking was not merely to serve as their "keeper," as he related to Prange:

In 1938, Fuchida "frequently led bombing missions over China, mostly in the Hankow region…the navy considered this assignment temporary duty and did not cancel out his Naval Staff College experience. Therefore, as his records show, he completed his studies on 15 September 1938."

Not so curiously, perhaps, given the "spin" which has so often been noted in "his" work, Prange's logic, here, is topsy-turvy. A more logical assessment would have been that, 'his records show that he completed his studies on 15 September 1938; *therefore*, the navy apparently considered this assignment temporary duty…etc..' Faced with Fuchida's *personal* recollections, which placed him in combat in China, and *official* records, which showed him to be attending the Naval Staff College during the same period, Prange had simply turned logic around to cover the lack of an official record of Fuchida's temporary duty. It appears, therefore, that assignments to the Naval Staff College offered a "cover" for special operations, and Fuchida's service record seems to indicate such a cover:

Fuchida had been assigned to the college on December 1, 1936, but had left for Nanking in late December, 1937, completing his staff college "studies" in mid-September of 1938 and transferring to Sasebo Naval Base on December 1, 1938. Perhaps because Fuchida realized that he had already talked too much, Prange could add only a few details as to his service in China. "Fuchida had little to say about his experience in China, beyond the fact that he had served at Shanghai and on Hainan Island as well as at Canton…" (where the *Kweilin* had been shot down).

Fuchida's association with the Naval Staff College had encompassed five events: the Earhart disappearance, the *Panay* sinking, the *Samoan Clipper* bombing, the *Hawaii Clipper* hi-jacking and the *Kweilin* shoot-down. But, for the last three of these events, Fuchida was nowhere near the Staff College.

As to Fuchida's determined and ambitious friend, Lt. Cdr. Minoru Genda, Prange notes Fuchida as recalling that, "Genda had come to the Staff College approximately one year before Fuchida, and for some seven months they served together once more." That is, Genda had come to the Staff College around December of 1935, and, because the two had parted company "some seven months" after Fuchida's arrival (on December 1, 1936), Genda had thus departed on, *or about*, July 1, 1937. This recollection placed Genda and Fuchida together until about the time of Earhart's disappearance on July 2, 1937. If Fuchida could have been released from the college for *undocumented* "temporary duty" in December of 1937 (at Nanking), so, one imagines, could Genda have been released for "temporary duty," in late June, 1937.

As to what these officers required, in the way of the means to achieve their dreams of glory: a suitable fighter, an attack plan, the operational tools and a rallying symbol, or "icon;" these could have been provided, in 1937, by one person: Amelia Earhart, who disappeared en route from Lae at about the same time that Genda left the Naval Staff College. This is not to suggest that Genda *hi-jacked* Earhart, but that he may well have been connected to her disappearance.

In 1985, the author observed two aircraft, curiously displayed at the National Air and Space Museum, in Washington, D.C. On the ground floor of the southwest display area, was a Mitsubishi *A6M5*, a late mark of the famous *Zero*. And on the upper floor of the northeast display area was Howard Hughes' *H-1*, the *American Racer*—as far from the *Zero* as was possible within the museum. The *Zero* was displayed on the floor and was viewed from a platform above the aircraft, while the *Racer* was suspended from the ceiling and could only be viewed from below. Visual comparisons were difficult, perhaps deliberately so, but it appeared quite plausible, as several aviation writers had already asserted, that the *Zero* might well have been a military version of the *Racer*. (The display setup has since been changed.)

Following Hughes' ill-fated attempt at a new speed record, there were three people who knew, first-hand, the military potential of Hughes' *Racer*. Howard Hughes knew: he had completed a successful run and had only to complete a second, to claim the world speed record, but the *Racer* crashed during the second run, sustaining heavy damage. Paul Mantz knew: Mantz was Earhart's technical advisor, but that day, he was flying the course as an observer for Hughes. And Amelia Earhart knew: she was the timekeeper, riding with Paul Mantz in the chase plane, and she had clocked the *Racer*'s speed during its first run. She could not but have known that the *Racer* had the potential to be transformed into a superb fighter.

As to a plan for an air attack on the U.S. Navy, not in Japanese home waters, as the Imperial Navy had long planned, but in the U.S. Navy's own waters at Pearl Harbor, a plan had already been laid out in 1924, and not by the Japanese, but by General "Billy" Mitchell, as an hypothetical appendix to his 1924 *Pacific Report*.

Comparing the actual attack on Pearl Harbor to Mitchell's plan for such an attack leaves one with an uncanny feeling that Yamamoto owed more to Billy Mitchell than to his own strategists. Mitchell's plan was especially well suited to a Japanese surprise attack because, even as Mitchell had endured the disgrace of a court martial, his theories had been equally discredited, to a large degree. Japan could thus adapt and implement his Pearl Harbor attack plan without fear of encountering any specific American defense against it. His report had been filed away by the U.S. Army, but authors generally retain copies of their own work, and General Mitchell's publisher was George P. Putnam, Amelia Earhart's husband.

As to a means of mounting long range naval air operations, that is, an efficient system of air navigation, there were few experts quite so valuable as former PAA Chief Navigator, Fred Noonan, whom Earhart had insisted upon employing as her navigator (despite his self-evident alcoholism and the advice of her friends and advisors). Fred Noonan became "available" to the Japanese in early July of 1937, in the southeastern Marshall Islands.

As to a graphic "icon," to provide symbolic support for a Pacific War against the U.S., Earhart, herself, was ideal: As the helpless "victim" of manipulation by her own nation, which had insisted on sending an unarmed, untrained female "pacifist" on a spy flight, she would appear as an undeniable example of U.S. deceit and treachery toward Japan. As an American public figure with political contacts in high places, who counted, among her friends, the American President's wife, Eleanor Roosevelt, Amelia Earhart was an undeniably valuable informant. As a *pilot* of courage, if not of great skill, she embodied virtues which the Japanese could appreciate. As a *woman* of courage, she illustrated, by contrast, the apparent spiritual bankruptcy of American men, who had sent a *woman* to do a job for which they seemed to lack both the skill and the spirit. And, although she was a woman in their male-dominated society, she was a *foreign* woman, who appeared to pose no threat to Japanese culture, or to the codes of Japanese officers. (Unhappily, for Japan, however, she was also a "Russian" woman.)

As to whether she would have been trusted, she had, after all, defected. She had brought gifts of documented knowledge, and she had even sold out Fred Noonan, her navigator. But her ace-in-the-hole may have been a bit more primitive than her defection "gifts:" An incipient red-head, who had made a redhead's way among men since her 'teens, Earhart, at forty, was still a woman worth bedding, and her new Japanese associates were hot-shot navy pilots—*tail-hookers all*.

The extent of the Earhart saga demands far more attention than the scope of this work can provide. The author has begun a detailed work, *Twice A Traitor*, which will shed light on Earhart's betrayal of both America *and* Japan, in a bold and successful attempt to save the Soviet Union from a two-front war. For now, a few points will have to suffice:

> That Earhart was, in her time, an active, outspoken and determined radical socialist is indisputable. That she crossed the line into Soviet Communism is made clear in certain of her letters, which have been published.

> That Earhart knew her exact location, *and reported it*, before ditching at Mili Atoll, in the Marshalls, has been proven by analysis of her final messages (*Appendix, Section 3*).

> That there were *three* persons aboard her plane, on the day that she disappeared, is indicated by reports of a retired Coast Guard officer, that, during her last transmission, an *unrecognized* voice "shrieked" something, after which her radio was simply turned off. Her last transmission was so troubling that USCGC *Itasca*'s radio logs were falsified.

> That she ditched in the "eastern fringes of the Marshalls" (G. P. Putnam's phrase)—not in the Phoenix group—has been proven by an analysis of Pan American RDF reports, which escaped destruction (as ordered by Naval Intelligence) only through the courage of a PAA secretary, Ellen Belotti.

> That Earhart lived within the jurisdiction of Japan from mid-July, 1937, until 1945, is validated by a preponderance of eye-witness reports, originating in widely separated areas of the Empire and privately gathered in the years since the war's end. These include two remarkably similar reports, from separate locations, each indicating that Noonan had been held under close guard, while Earhart remained at liberty. There is even a report of their executions—attributed to Genda!

Earhart was certainly the *Russian woman* (a political reference), reportedly executed at Saipan before the war, and also the *foreign woman*, whose affair with Yamamoto had angered the young officers of his command. Earhart had the *motive*, the *means* and the *opportunity* to provide the Japanese with a complete Pacific War "package"—complete, that is, except for an engine to power their *Racer*-type fighter, the *Zero*. To obtain such an engine, Japanese naval aviators would eventually hi-jack the *Hawaii Clipper*.

A profile of *Hawaii Clipper*'s hi-jackers is not difficult to outline. Technical necessity demanded that they be airmen, themselves, and security concerns demanded that they be on active service with the Imperial Japanese Navy. No enlisted man could be entrusted with such a sensitive task and no officers below middle-rank would have been able to coordinate the naval air support, which the mission required. (Those readers who doubt that middle-ranked naval officers, with some protection from more senior officers, could have promulgated such a complex operation, are invited to compare the Clipper hi-jacking to the Iran/Contra operations of the 1980's, and Minoru Genda to Oliver North.)

Because the three American aircraft losses (Earhart in July of 1937, *Samoan Clipper,* in January of 1938, and *Hawaii Clipper,* in July of 1938) all appear to have been part of a progressively sophisticated development pattern, it seems likely that the lead hi-jacker may have planned and promulgated all three incidents, including Earhart's defection.

Since the *modus operandi* in all these cases appears to have involved a hi-jacker stowing away aboard a civilian aircraft in a Western-controlled port (that is, Lae, Pago Pago and Sumay), the lead hi-jacker had to present a western appearance in stature and features. Of the Japanese naval aviators of the period, one man could easily have fit this profile, and that man was *Minoru Genda*.

2. THE HI-JACKING

1. OVERVIEW: The Clipper Hi-jacking of 1938—and the Ultimate M.I.A.'s

At 5:39am, in the blue-gray dawn of July 29, 1938, Pan American Airways' trans-Pacific flying boat, *Hawaii Clipper*, one of but three hard-pressed Martin *M-130*'s, cast off from the terminal slip at Sumay, Guam. She was four days' flying out of Alameda, bound for Cavite on Manila Bay; then on to China, to end Trip 229 at Hong Kong, seven days out. Embarked aboard her that Friday morning were nine crew, six passengers, a moderate load of 2550 gallons of gasoline and 1138 pounds of cargo—and two stowaways. At 6:08am, just before sunrise, after what some observers thought was an unusually long take-off, she lifted from the waters of Apra Harbor, leaving Guam behind—forever.

She settled quickly into the routine of flight, and, as her passengers steeled themselves to another twelve tedious hours in the air, the expanded crew began the watch cycle, each logging time for growth with the flourishing airline. The Flight Radio Officer began the departure sequence, sending back reports, in Morse, every half-hour, in return for course confirmation from Sumay's Radio Direction Finder. Beyond mid-flight, as the Philippine stations, Radio Panay and the Makati RDF, signed on for the arrival sequence at Cavite, Radio Sumay would stand down and, signal reception permitting, serve as a monitor.

Not long after takeoff, in the ennui born of droning engines and high altitude, the two stowaways made their move. Undaunted by the odds against them, their daring plan was to convince PAA, through a brilliant trick of radio-deception, that the Clipper had come to grief in flight—on course to Cavite—even as she flew on to Ulithi and, thence, to Truk, Japan's vaunted *Gibraltar of the Pacific*. If they failed, then *Hawaii Clipper* would be intercepted farther along her route, and they, too, would die, along with the Americans.

Hi-jacked!—not that it wasn't an admitted risk: Japan was then at war with China, and PAA's Clippers had pierced the Japanese naval buffer in the Mandates. First Officer Mark Walker, a Navy Reserve carrier pilot, had mastered the Pacific, along with others, at no risk—or cost—to the U.S. Navy. And passenger Ted Wyman, also a Navy Reserve officer, was off to China as a vice president (for export) of Curtiss-Wright, whose *Hawks* already fought for China and whose *P-40's* would later fight with the *Flying Tigers*. It was the dawn of the Pacific War, and Pan American Clippers flew the dawn patrol.

Yet this was no act of war, but an act of *piracy*, committed on and over the high seas against a duly registered American merchant vessel—wings and all! As with all acts of piracy, the victims were a liability, and none would live to tell the tale, but these "pirates" were no petty buccaneers serving under the Jolly Roger. Few but officers of the Imperial Japanese Navy would have dared so much. Fewer still would have jeopardized both flag and nation but fanatic, middle-rank aviators of the Imperial Navy's hostile *Fleet Faction*, whose flag was Togo's provocative *Rising Sun with Rays* and whose nation would soon be forced to conspire in this: the first, and, perhaps the *worst*, of all airline hi-jackings.

And all for *what*? For a suitcase, filled with U.S. Gold Certificates? For an end to a study of Asian "germs," carried to America on the winds? For Japanese air superiority, assured over China? For an engine, to be copied for the *Zero*? For a new war plan, altered from a military *land war* against the Soviet Union, to a naval *air war* against the United States? For *glory*? Oh, yes—for *all of these*.

But, if the Japanese "pirates" were brilliant, the American navigator was *inspired*.

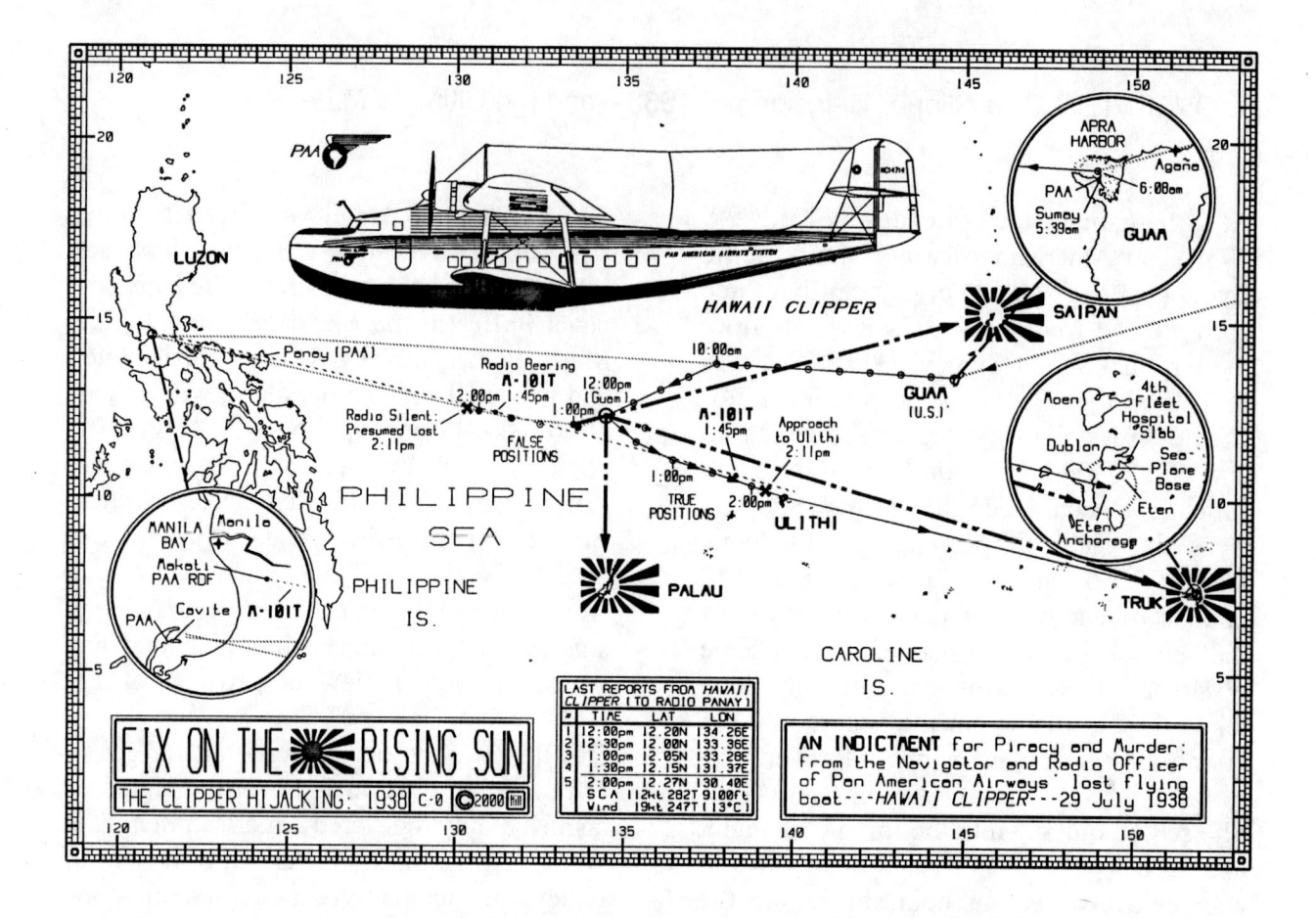

Compelled to plot two routes, one true, to Ulithi, and one false, to Cavite (both routes indistinguishable at the Makati RDF), the Clipper's Second Officer, George M. Davis, reconfigured three of his last four precisely calculated—but *false*—positions, to pass the word to anyone who might re-plot the route. Substituted for the Japanese-approved false positions by FRO William McCarty, as he transmitted the deceptive flight reports, the reconfigured positions were at odds with other known flight parameters and clearly invalid as elements of Davis' dead-reckoned navigation. Instead, they established three intersecting lines, each fixed precisely on a major seaplane landing area in the Japanese Mandates. By simple extension from general practices of standard navigation, these three lines of position constituted a *running fix*, of sorts, on three seaplane bases of the Imperial Japanese Navy, providing a clear message, for anyone who *got* the message, as to what had happened, who was responsible and just where the big PAA flying boat was actually headed—literally, a *Fix on the Rising Sun*.

Yet, even if Pan American Airways and the U.S. Navy seemed *not* to "get the message," it still exists for us to read, today. And of the ultimate fate of these travelers, a trace may yet remain: on Dublon Island, at Truk Atoll in the Caroline Islands (now the Federated States of Micronesia), lies a weathered slab of concrete, poured in the late summer of 1938—all that still remains of the Imperial Navy's infamous Fourth Fleet hospital. The Japanese consider this site of medical war crimes atrocities to be a shrine, and many unknowing tourists have stood upon it and contemplated the countless thousands who died in the bitter Pacific War. But there is nothing to remind the visitors of *fifteen* who were murdered at the dawn of that war and *who lie beneath their very feet,* sealed in the concrete, it is said, "with no marks of death upon them," truly—*the ultimate M.I.A.'s*.

36

2. A WESTERN PACIFIC GAME-BOARD: The Philippine Sea Area, 1938

A last bastion of colonialism, the Philippine Sea area, with its numerous island groups, became a focus of greed and conflict among many newly "emerging" nations in the late nineteenth century. For Japan and the U.S., newcomers to the colonial game, the area contained the only potential colonies for either, colonies that could provide a key to controlling the wealth of Asia.

The United States, prompted, perhaps, by a strong German colonial presence within the Caroline, Marianas and Marshall Islands, evicted Spain from the Philippines and from Guam in 1898, during the Spanish-American War. The acquisition of the Hawaiian Island chain accompanied the Philippine victory and the entire string of American territories, including Guam, Wake and Midway Islands, created a guarded route to Asia. Japan was incensed by this American colonial thrust in the Pacific (and especially the acquisition of Hawaii), but, with Germany a world away, the U.S. enjoyed a useful time in which to develop its Pacific colonies. The Pacific sea lanes to Manila and the southern Philippines quickly became an American highway to Asia. (In 1935, in partnership with the U.S. Navy, Pan American Airways would bring trans-Pacific air service to the area—and add a new sense of immediacy to the game.)

After 1898, American acquisitions had left no territory in the western Pacific for further colonialism, and Japan became fearful that her own brand-new navy might soon find itself all dressed up—but with no place to go. So, in 1903, Japan turned its eyes to the north and, in 1904, found itself at war with Russia, in an attempt to grab all of Sakhalin Island and the Kuriles. Under Admiral Togo, the John Paul Jones of the modern Japanese Navy, Japan won a decisive naval victory over the Russian fleet at Tsushima Straits, and it was in that battle that the world first came to see that striking battle ensign, the *Rising Sun with Rays* (Admiral Togo's own design), under which the Imperial Japanese Navy would later attack the United States and officially initiate the Pacific War.

Togo's naval achievement became a hollow victory when President Theodore Roosevelt, as "peacemaker," dictated the terms of the 1905 Treaty of Portsmouth, by which Japan settled for limited power in Manchuria, half of Sakhalin Island and a part of the Kuriles (including Etorofu, from whose Hittokapu Bay, the "attack force"—the *Kido Butai*—departed for Pearl Harbor in 1941). In the wake of this offensive (to the Japanese) treaty, a powerful faction was born in the Japanese Navy, the so-called *Fleet Faction*, which rejected diplomacy and believed that Japan's ultimate destiny should depend upon conquest by sea, guided by Togo's spirit and conducted under his bold new colors.

World War I opened up new possibilities for Japan, and so, the Japanese turned their eyes southeastward to the Caroline, Marianas and Marshall Islands. In order to establish a potential claim to German colonial holdings in the islands, Japan allied itself with France and Britain—in Europe's *Great War*.

After WWI, there came both the creation of Wilson's League of Nations—and the U.S. failure to become a part of that organization. The Japanese convinced the League that the bulk of Germany's Pacific island colonies should be mandated to Japan (with Britain receiving control of German colonies in New Guinea and the Bismark Archipelago). The U.S. was in no position to counter this move and so Japan, with the stroke of a pen, suddenly found itself to be a major colonial power, strategically athwart the U.S. supply

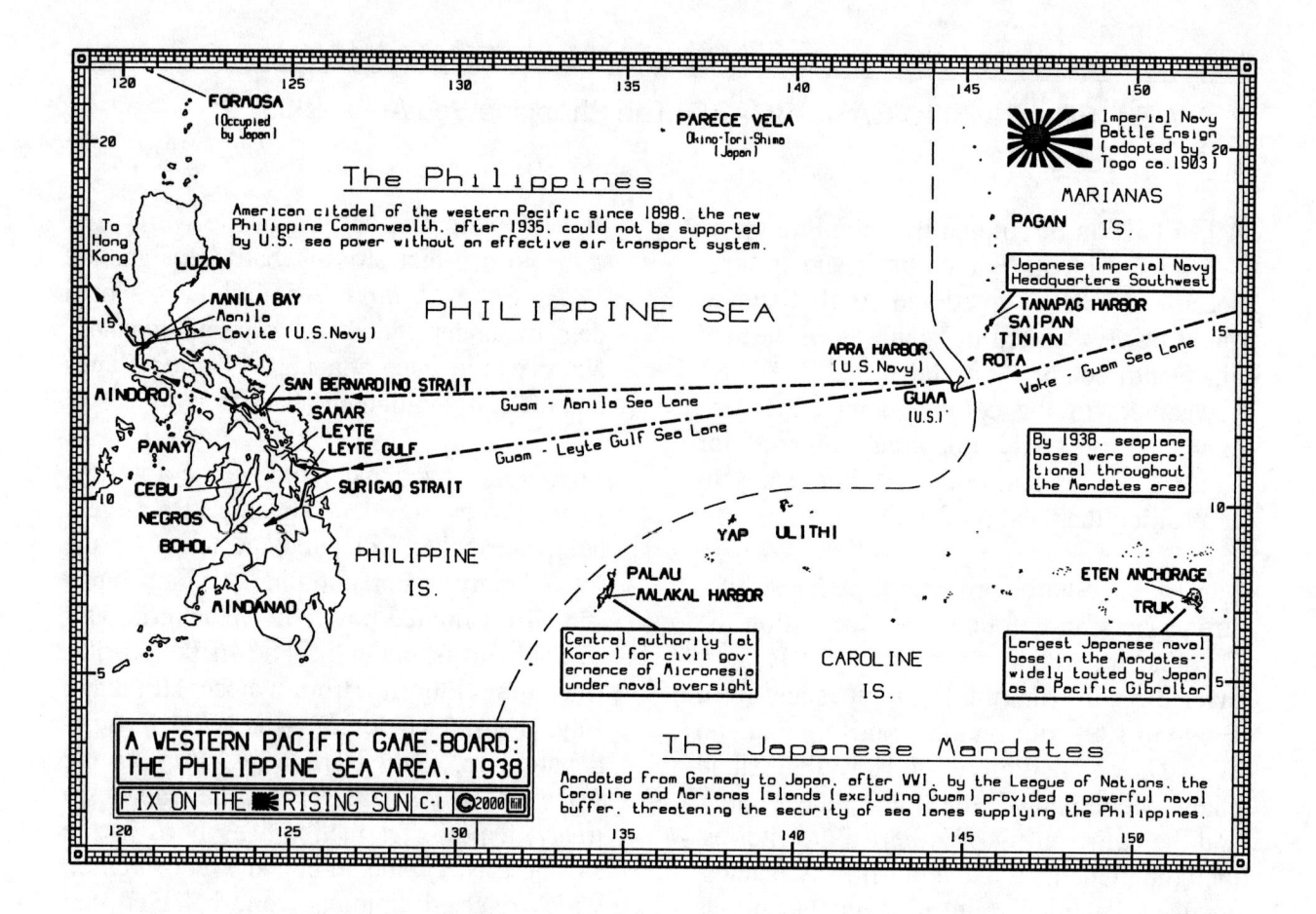

A WESTERN PACIFIC GAME-BOARD: THE PHILIPPINE SEA AREA, 1938

The Philippines
American citadel of the western Pacific since 1898, the new Philippine Commonwealth, after 1935, could not be supported by U.S. sea power without an effective air transport system.

The Japanese Mandates
Mandated from Germany to Japan, after WWI, by the League of Nations, the Caroline and Marianas Islands (excluding Guam) provided a powerful naval buffer, threatening the security of sea lanes supplying the Philippines.

Map labels include: FORMOSA (Occupied by Japan), PARECE VELA (Okino-Tori-Shima, Japan), Imperial Navy Battle Ensign (adopted by Togo ca. 1903), MARIANAS, PAGAN IS., Japanese Imperial Navy Headquarters Southwest, TANAPAG HARBOR, SAIPAN, TINIAN, ROTA, APRA HARBOR (U.S.Navy), GUAM (U.S.), Wake-Guam Sea Lane, Guam-Manila Sea Lane, Guam-Leyte Gulf Sea Lane, To Hong Kong, LUZON, MANILA BAY, Manila, Cavite (U.S.Navy), MINDORO, SAN BERNARDINO STRAIT, SAMAR, LEYTE, LEYTE GULF, PANAY, CEBU, SURIGAO STRAIT, NEGROS, BOHOL, PHILIPPINE IS., MINDANAO, PHILIPPINE SEA, YAP, ULITHI, PALAU, MALAKAL HARBOR, ETEN ANCHORAGE, TRUK, CAROLINE IS.

Notes on map: "By 1938, seaplane bases were operational throughout the Mandates area." "Central authority (at Koror) for civil governance of Micronesia under naval oversight." "Largest Japanese naval base in the Mandates—widely touted by Japan as a Pacific Gibraltar."

FIX ON THE RISING SUN C-1 ©2000

---

routes to the Philippines. Japan's diplomatic acquisition of these island groups produced a second faction within the Japanese Navy: the *Treaty Faction*, which sought to exploit diplomacy as a viable, cost-effective means of pursuing Japan's Pacific destiny.

By 1938, Japan's army had become bogged down in a land war in China and had failed to achieve the strategic position required for a planned invasion of Soviet Siberia. Japan's navy, however, while relegated to providing coastal air support for the stagnating army forces in China, was in no way demoralized. The naval forces of Japan's Island Empire had been brought to such an exalted state of readiness that American naval and military planners had already decided to write off the Philippines in the event of open war. On the *diplomatic* front, then, the U.S. was forced to operate with some caution, for, if Siberia was in doubt as a Japanese target in 1938, the Western Pacific certainly was not.

Japan centralized its control of the Mandated Islands by creating a "front door" at Koror, in the Palau Islands, which regulated both civil and commercial access to the Mandates under the auspices of the Imperial Navy. At Saipan, the Navy situated its headquarters, on the fringes of the Mandates, but securely isolated from Tokyo. And at Truk, they had already (in 1915) created a large naval base, touted as the "Gibraltar of the Pacific"—and virtually immune to a naval surface attack.

Restricted by treaties and by economics, the Imperial Navy built few facilities for land-based aircraft. Instead, they concentrated on large flying boats and the inexpensive bases needed to service such aircraft. And in 1936, Kawanishi began producing the *H6K* flying boat (a copy of the American Sikorsky *S-42*, which was used by Pan American Airways), and, throughout the Island Empire, these flying boats began providing air service for command, supply and intelligence missions.

3. PAA AIR ROUTES TO THE ORIENT: Bridges over a Sea of Turmoil

Pan American Airways' expensive decision to open air routes from the U.S. to Asia was certainly not undertaken simply to satisfy a commercial need. PAA enjoyed a very close association with the U.S. Government and had been founded, initially, at the insistence of the U.S. Army, to preclude any American dependence upon a German airmail carrier in the Caribbean area. In the Pacific, PAA's use of secure U.S. Navy (and Marine) bases at Pearl Harbor, Midway, Wake, Guam and Cavite leave little doubt that Pacific Clipper service was established, in 1935, in the clear interest of American national security.

PAA had a number of problems to solve in opening the Pacific routes: Pacific weather could be both severe and unpredictable, and, except for Midway Island and Wake Island, the terminal sites were located near rugged mountainous terrain, thus making runway construction expensive and difficult and the landing approaches dangerous (especially through the overcast). Radar did not yet exist and long-range radio communications were limited by atmospherics, common to the Pacific, and by the lack of suitable radio equipment. Aerial navigation was still an undeveloped art, and although the routes to be covered were much longer, per leg, than had previously been flown by regular air carriers, several of the terminals were to be located on small, isolated Pacific islands.

PAA's long experience with flying boats in the Caribbean certainly dictated the type of aircraft, which it would use in the Pacific. But the older Sikorsky *S-42*'s were too small for cost-effective trans-Pacific service (other than for survey work and for New Zealand mail service), and so PAA contracted with the Glenn Martin Company, of Baltimore, Maryland, to build three large flying boats, the distinctive *M-130*'s, for the Pacific route.

PAA also created a new support company, PAMSCO, which designed and built the electronic equipment required for long range service. These included radio transmitters, receivers and radio direction finders for use on the ground and in the air. The RDF's were of two types, the traditional loop-type and a sophisticated multi-antenna type, the Adcock, which was adapted from a British design. The Adcock type was quite sensitive and, properly calibrated, could provide valid radio bearings, often with an accuracy of less than a degree of arc, on high frequency signals originating a thousand miles away.

The development of the PAA terminal bases included radio and RDF stations, terminals, hangars, maintenance shops, prefabricated hotels (for Midway and Wake) and even a narrow gauge railway (for Wake).

For navigation, Pan American Airways had the services of one of the best air navigators in the world, Frederick J. Noonan. A former windjammer officer (with required Master's papers, but no pilot's license), Noonan had served as a PAA inspector in the Caribbean and had made great strides in development of Great Circle navigation. For the much lower latitudes of the western Pacific, Great Circle navigation was never required (except for RDF bearings), but Noonan still faced the problem of finding isolated islands, after flying blind for up to fifteen hours through unpredictable Pacific weather. For safety, he insisted that all of PAA's navigators rely entirely upon conventional navigation, using radio only as a valued check on their work. And, contrary to the risky, but common later practice, that of "riding" a radio "beam" on a direct route, Noonan preferred flying to the left or right of course, to plot running fixes and to ensure a confident final turn toward a destination—the Single Line Approach.

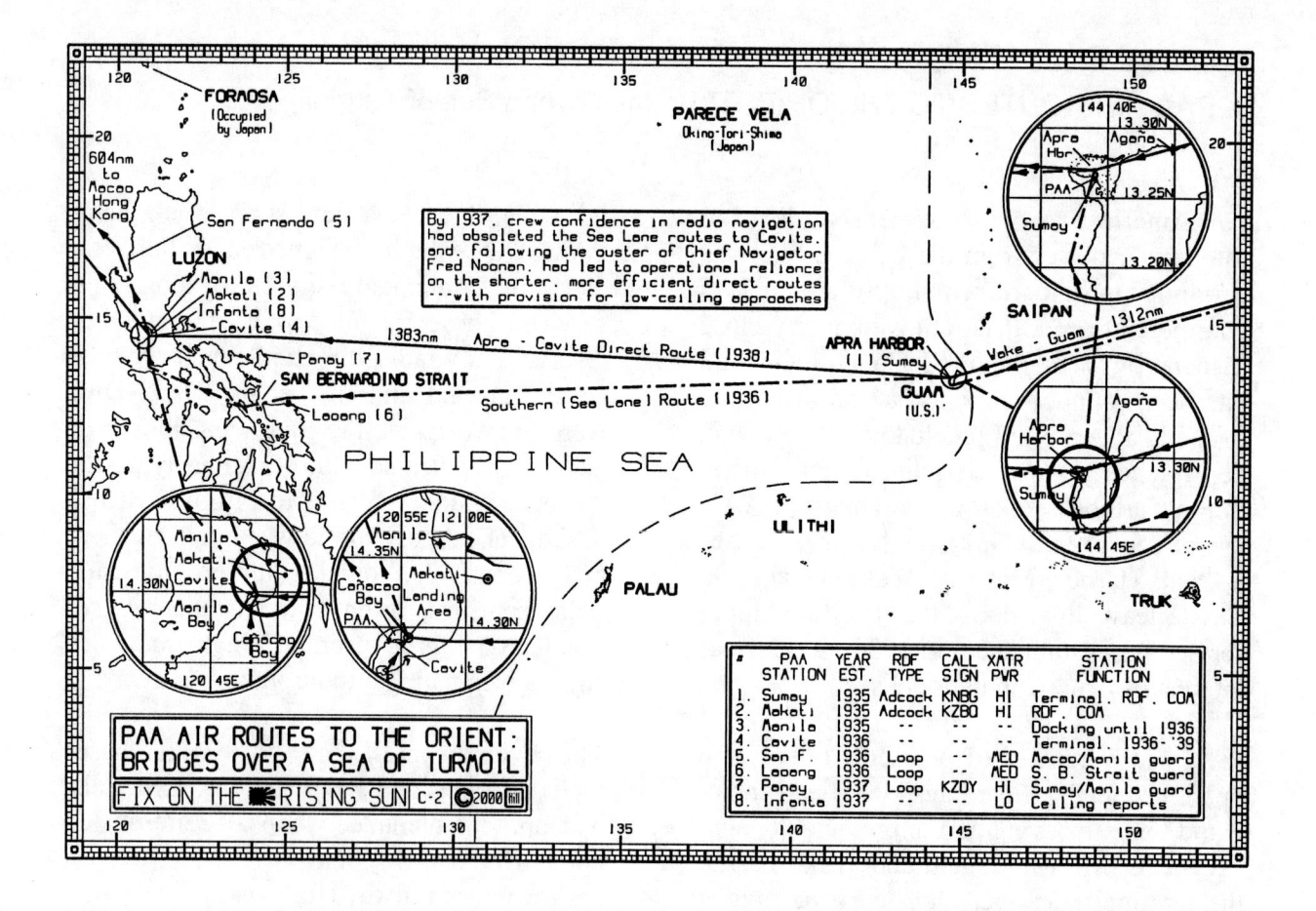

The original routes on which PAA began its Pacific air service closely followed maritime sea lanes, such that the Clippers could have "sailed" the entire route, from California to Hong Kong. More realistically, they could, if disabled en route, count on the assistance of passing vessels. In fact, PAA Clippers were legitimate maritime vessels and were always described as "flying boats" and never as "seaplanes." Each Clipper carried its own name on the fuselage, and, in addition to its national aircraft registry number, each had an American flag painted on its hull. The Captains were all qualified master mariners, and the crews always referred to the cockpit as the "bridge." Yet the Clippers were also aircraft, of course, and, as radio navigation eventually proved to be reliable, political pressure grew among senior flying officers to replace Frederick Noonan, to discontinue reliance upon conventional navigation and to abandon the original routes along the longer and less efficient maritime sea-lanes.

As is well known about Fred Noonan, he disappeared with Amelia Earhart during her famous round-the-world flight on July 2, of 1937, supposedly after having been fired from PAA for alcoholism. What is not well known is that alcoholism only provided a convenient excuse to force Noonan out as PAA's Chief Navigator: the fact that PAA sea-lane routes became obsolete at about the same time appears to indicate that Noonan was, indeed, replaced to permit the shorter direct routes to be used. In fact, few retired PAA pilots, who flew the Pacific no earlier than 1938, remember ever having used the then-obsolete sea-lane route through the San Bernardino Strait. Noonan's concerns about safety had not been completely unfounded, however, and, in 1937 (as shown in Chart C-2, above), PAA established approach bases at Panay (high power radio and loop RDF) and at Infanta (low power radio and cloud ceiling observatory), to support the riskier direct route over the mountains to Cavite.

4. GUAM: PAA Operations in the Area of Guam

In 1935, as a result of the grand decision to establish a trans-Pacific air route to Asia, PAA began acquiring sites upon which to build their terminal and radio facilities. At Guam, PAA was permitted the use of a former U.S. Marine base, belonging to the U.S. Navy and located at Sumay, adjacent to Apra Harbor. The facility was modified to suit PAA's special needs and included a terminal, docking slip, hangar and aircraft repair area, as well as an Adcock-type Radio Direction Finder (with its extensive antenna field) and a radio communication facility. Unlike other PAA terminals, the Sumay site was sufficient to contain all of the required radio equipment and antenna fields.

Early in the trans-Pacific program, Clipper flights arriving from Wake Island, over 1300 nautical miles to the east-northeast of Guam, approached the island from the south, flew north along the west coast to Apra Harbor, then turned east-northeast toward Agana, the capital city of Guam. Once over Agana, the Clipper would turn back again toward Apra Harbor, approaching the landing area from the east in order to land into the prevailing westerly winds (that is, the winds from the west). This roundabout approach to Guam was required for three reasons:

First (and following Fred Noonan's normal practice on long flights to very small land masses), by flying to the left of the nominal course from Wake Island, a navigator could always be confident that Guam would lie to the right—once it had flown the estimated distance, or Distance Made Good (DMG). For instance, if the Clipper flew directly to Guam and, upon achieving Distance Made Good, found nothing but empty sea below, there would be no valid way of determining whether to seek Guam to the right or the left of the course. However, if the Clipper were deliberately flown to the left of the Direct Route, that is, on the Southern Route, and achieved DMG far to the left of that course (at 13.00N, 144.45E, for example) there would still be nothing but empty sea below. However, rather than simply being lost, the navigator could be confident that Guam lay somewhere to the right, having held a course calculated to be far enough to the left to ensure that, with normal errors (for a certain flight and flight crew), the island *must* lie to the right. If, on the other hand, the Clipper arrived far to the right of the Southern Route (at 13.30N, 144.45E, for example) some part of Guam would lie directly beneath the Clipper, and, if the island were not under cloud cover (another serious problem), it would be clearly visible for a final approach.

This brings up the second reason for the Southern Route, which has to do with both the weather and the mountainous terrain of Guam. If, upon achieving DMG, a Clipper found itself above unbroken cloud cover, with no radio by which to learn of clearance between the mountain tops and the bottom of the clouds, the Southern Route permitted a safe blind descent to the *left* of the route, followed by a low level approach to Guam *under* the overcast. Since the highest point on Guam is only about 1300 feet above sea level, the hazard, at Guam, was certainly less than at Manila, but there was definitely a requirement, early on in the program, for navigational procedures that relied neither upon unproved radio navigation systems, nor even upon radio communication at all. Fred Noonan's reliance upon conventional navigation and Distance Made Good (rather than the far less predictable Course Made Good) was sensible, not only for the early days of trans-Pacific flying, but even for the war years, when the *Philippine Clipper* was lost in the mountains north of San Francisco.

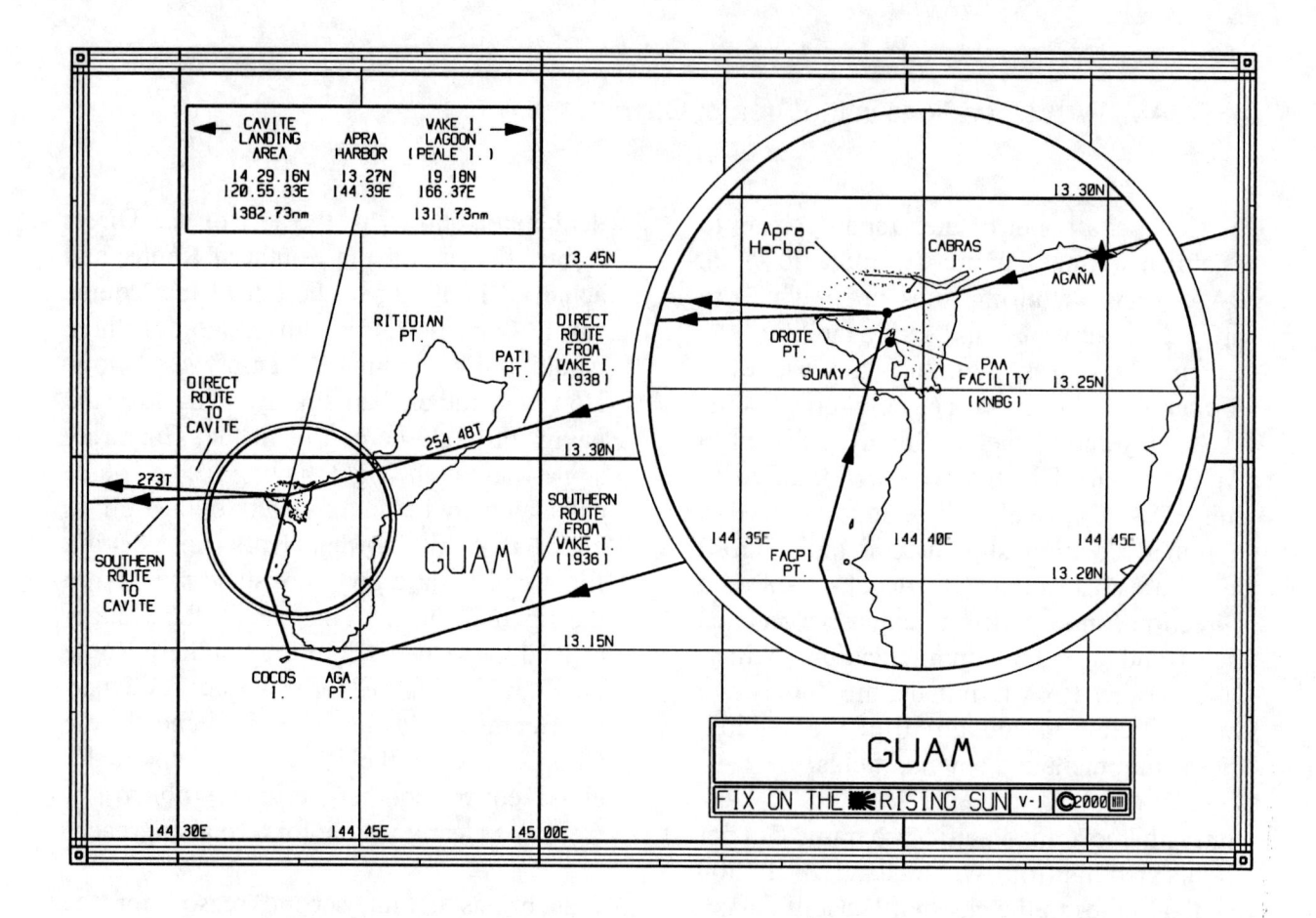

The third reason had to do with the islands of Rota, Tinian and Saipan, which lay to the northeast of Guam and which were under tight Japanese naval control. The Southern Route to Guam ensured that no inadvertent over-flight of any of these Japanese-held islands might occur. Although some former PAA personnel have suggested that a few deliberate over-flights did occur after 1938, in the interest of intelligence gathering, it was imperative, in the beginning, not to add Japanese political outrage to the technical problems that had to be resolved.

Following Fred Noonan's ouster from PAA, the Direct Route was generally followed, and few surviving PAA flight crewmen, in the early 1990's, recalled having flown the Southern Route into Guam. As it happened, there were never any recorded instances in which the Direct Route to Guam placed a Clipper in jeopardy. However, some old hands from PAA's Pacific days might still recall the tragedy of the *Philippine Clipper*, which crashed, far north of San Francisco, in 1942. Mistaken as to her position above the overcast (and forbidden to fly to an alternate base), she descended into a mountainside, where it took several weeks to locate her.

It will be understood that concerns regarding the simpler Direct Route versus the far safer Southern Route were generally applicable only during inclement weather and/or under the threat of radio failure. Noonan's caution, during the early years of trans-Pacific PAA air service, might seem excessive to airmen, today, who are familiar with present-day communication reliability, but it was not at all excessive, either for Noonan's time or for the later years of the 1930's. In fact, as will be shown, it was the unquestioned faith in PAA's excellent record of communication and RDF reliability—far too soon after that reliability had been proven—which finally led to the loss of *Hawaii Clipper*.

5. 2 CLIPPER APPROACHES TO CAVITE: 1 as by Sea—1 out of the Blue

If the relatively low mountains of Guam posed a minor hazard to aircraft, the rugged terrain of the Philippines was—and is—a hazard of major proportions. PAA's own concerns can be seen in the development of its radio facilities during the early years of trans-Pacific service:

A. 1935, Makati: The original radio station established in Luzon by PAA, Makati was intended to support an approach to Manila in any weather. However, although the Adcock RDF could provide adequate radio bearings for much of the western half of the Guam-to-Manila leg, it was overly sensitive (even to roaming animals) and was not sufficiently well-placed to support a blind approach to San Bernardino Strait, which, during heavy weather, had to made at low altitude.

B. 1936, Laoang: A small station at Laoang was established when the need for a radio "beacon" near San Bernardino Strait became apparent. Equipped with a medium-power transmitter and the loop-type RDF, Laoang could provide either usable radio bearings on the incoming signal from an approaching Clipper or a steady outgoing signal on which a Clipper could take its own bearings. When foul weather settled onto the Philippines, a Clipper could fly the Southern Route direct to the San Bernardino Strait, then drop down and continue to Manila at low altitude, just below the 2000 to 3000 foot cloud ceiling. Under these poor conditions, "navigation," was replaced by "pilotage," which depended upon the sightings of visual landmarks. This was Fred Noonan's safe Southern Route.

C. 1937, Panay: After Noonan was ousted from PAA, opposition to the Direct Route to Manila quickly evaporated, but this route still needed a few modifications in order to provide security for foul-weather flying. A new radio station was established at Panay Island and equipped with a high-power transmitter and a loop RDF (as at Laoang). However, Panay's higher power enabled an operator to provide primary communication for the entire western half of the Guam-to-Manila leg, with Makati providing only the RDF services from its Adcock system.

Communication responsibility for all flights fell to the nearest station on any given leg: For example, for westbound flights, Sumay had responsibility for communications and bearings from take-off at Guam to the mid-point between Sumay and Panay, at which point Sumay simply monitored the flight communications, while Panay and Makati assumed primary responsibility for the rest of the flight. Because of the remote location of Radio Panay, only competent Filipino radio operators were employed by PAA and billeted at the station. On July 29, 1938, the day that the *Hawaii Clipper* disappeared, the operator at Panay was Edouard Fernandez. Communications, that day, were maintained (as always) by wireless telegraphy only, on 1638kcs (kHz). After Radio Panay assumed communication responsibility for the flight at 0200Z, however, Radio Sumay was not able to receive the Clipper's signals and was unable to provide the monitor function. It should be noted, with regard to telegraphy, that radio operators could generally identify each other, simply from the "swing" of an individual's familiar keying technique.

The route shown, passing just to the north of Panay, would have been the most nearly direct route to Manila, one which provided for accurate position plotting, using signals from Panay in conjunction with the Adcock RDF at Makati. However, it appears that some later Captains may have elected to "ride the beam" *directly* over Panay, which would not have permitted them to check on their true position by plotting a running fix against the offset signals from Panay. Such a practice would also hazard a flight if they missed the *Cone of Silence,* as they passed over Panay, in which case they would be unaware as to whether they were east or west of Panay. The route to the north, as shown, was clearly the best option for foul weather navigation along the Direct Route.

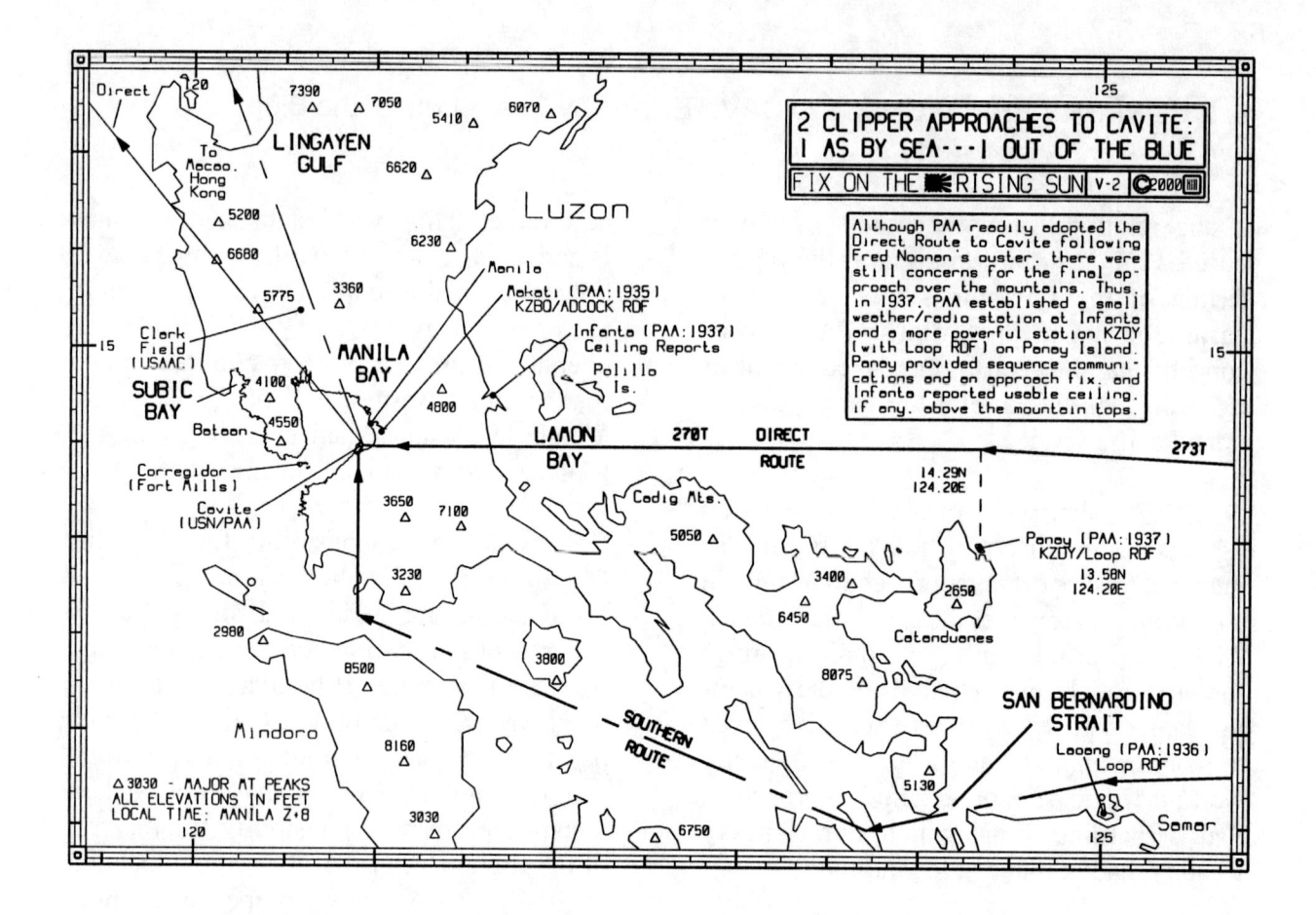

D. 1937, Infanta: At the same time that Panay was established, a small, low-power station was installed at Infanta, near Lamon Bay. Infanta's responsibility was to report cloud ceilings above the coastal mountains east of Manila. As Clippers approached on the Direct Route, Infanta would advise the aircraft of clearance between the tops of the mountains and the bottom of the cloud cover—if any existed. The day that *Hawaii Clipper* disappeared, the ceiling (above sea level) was 2500 feet, which means that *every*

mountain top, as shown on V-2, was above the bottom of the clouds; that is, *within* the clouds. Under such conditions, the Direct Route to Manila could be used only if the Captain could find a low mountain saddle well beneath the clouds, or, if the Captain was willing to climb high into the clouds and dead-reckon a descent into Manila Bay. If neither of these two possibilities seemed prudent, the Captain could land at Lamon Bay and wait for the cloud ceiling to rise— as was done on at least one occasion.

Generally, the Direct Route was dependable, the radios and RDF's proved reliable and the decision to discontinue the Southern Route appeared sound. But regardless of the route followed, it was clear that the 200 miles of mountains between Panay and Cavite were *always* to be avoided: On September 18, 1938, the *New York Times* quoted PAA officials as saying that "flying boats when in trouble avoid land and keep over the water."

6. MANILA BAY: PAA Operations in the Manila Area

Manila represented an entirely different challenge for PAA. The Manila Hotel, near the waterfront, provided accommodations which were quite suitable for the kind of passenger who could afford passage on a Clipper, but the crowded metropolitan and waterfront areas afforded little space for a terminal, the service facilities, or the Adcock RDF's extensive antenna field.

In the beginning, in 1935, service facilities were established at Cavite, and the RDF was installed at Makati, near Fort McKinley. But the Manila Hotel served as a terminal for the passengers, who were lightered out to the Clippers in a small motor launch, often in the early morning darkness. There were no dockside hangars near the hotel and security for the Clippers was unreliable. Rough sea conditions beyond the breakwater at Manila hampered landings and take-offs, and the Clippers often had to run a gauntlet of small craft in the crowded bay area. In 1935, the publicity value of PAA Clippers operating out of Manila, itself, may have made all the trouble worthwhile, but by 1936, security concerns had dictated a tighter operation.

As at Guam, the U.S. Navy assigned a Marine base to PAA, and, in 1936, PAA combined its terminal and service facilities in the southwest corner of Canacao Bay, at Cavite. The Adcock RDF remained out at Makati, but the sheltered landing area east of Canacao Bay provided a safer landing than that afforded by the landing area near the hotel. Unhappily, passengers had to endure a nearly twenty-mile drive to reach Manila, but the Clippers were no longer exposed to the hazards of the bay—or to the dangers of sabotage. Cavite's terminal was a temporary facility, and PAA planned a move, in 1939, from the isolated Marine base to the more secure Sangley Point Naval Air Station.

The original maritime route to Manila was over-long, as at Guam, and, by 1938, the Southern Route to Manila (through the San Bernardino Strait) was virtually obsolete. Cavite was, by then, clearly the center for PAA's Manila Clipper operations, although inbound flights always included a fly-over of Manila, as at Agana, Guam. This was a reasonable side-trip, if only to provide the paying passengers with a scenic view of the Philippine capital and to show the U.S. flag.

For the purposes of this work, however, it is unclear whether Manila or Cavite served as the official navigational arrival point for the flights from Guam. According to Captain Harry Canaday, who succeeded Noonan as Chief Navigator for PAA (and who may well have had a hand in his ouster), Manila remained the navigational destination point for flights from Guam, even after operations were moved to Cavite.

But, on the other hand, it is equally unclear whether Captain Canaday, who had flown for PAA in the Pacific as early as 1935, was still as close to operations beyond Alameda in 1938. It should also be noted that PAA's skippers were permitted—and exercised—a good measure of latitude in the choice of routes, even during flight. Finally, Cavite, lying further west than Manila, was, in fact, the actual flight terminus, by 1938.

This minor question, as to whether Manila or Cavite served as the navigational point for flights from Guam, becomes important in the analysis of *Hawaii Clipper*'s last flight, since Cavite, and not Manila, was her actual final destination—as was so clearly indicated in her final two reports. For the purposes of this navigational analysis, the author has made use of the Cavite Landing Area as the nominal arrival point.

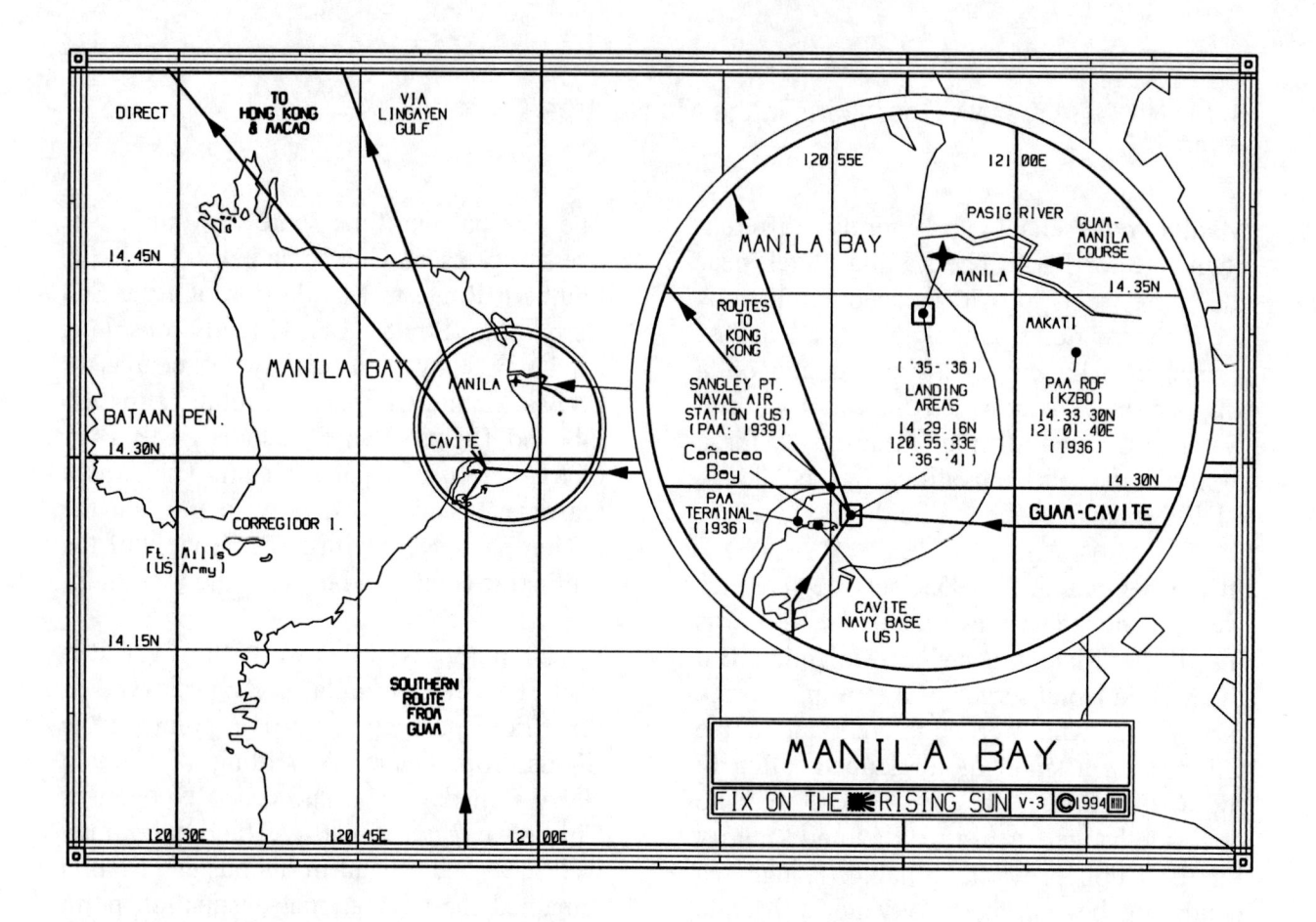

In any case, if Captain Canaday's statement, made in 1992, was true for all flights, then *Hawaii Clipper*'s unusual use of Cavite as a navigational arrival point would make her last flight all the more suspicious. If not, then the use of Cavite as a standard point of arrival in these charts has the effect of making the analysis easier to present and, perhaps, easier to understand. In either case, using Cavite as an arrival point has had no adverse effect on the analysis.

Manila Bay, of course, was not the western terminus for trans-Pacific "trips," as PAA termed them. There was an additional leg to Hong Kong and Macao, on the south China coast, initially provided by the *Hong Kong Clipper*. This Sikorsky *S-42B* not only saved each of the *M-130*'s at least three days out of twenty-one in every trans-Pacific trip cycle, but gave PAA a certain flexibility in arranging airline connections to Europe, via Imperial Airways, and to China, via PAA's troubled (and unpublicized) subsidiary, the China National Aviation Corporation.

In 1937, *Hong Kong Clipper* was withdrawn from service on the Hong Kong/Macao run, reconfigured with large internal fuel tanks, renamed *Samoan Clipper*, and assigned to inaugurate PAA's new airmail route to New Zealand—which was also a useful exercise in establishing an auxiliary air link to Asia, in the event that the Japanese ever closed the Philippine air or sea routes. But, on January 11, 1938, the *Samoan Clipper*, with Captain Ed Musick in command, exploded and then burned, reportedly while draining gasoline in preparation for an emergency landing in the harbor at Pago Pago, American Samoa.

This left the *M-130*'s to cover the extra leg to China, along with the trans-Pacific legs, thus increasing the overall traffic load by about fifteen percent and putting a severe strain on PAA's Pacific Division operations.

7. PREFLIGHT: PAA *Hawaii Clipper*—Trip 229 / Leg 5, July 29, 1938

There are three factors to be considered in planning a long flight, especially one over an expanse of ocean: (1) the air temperature, (2) the barometric pressure and (3) the Wind Vector, which includes both wind direction and speed. Temperature, and, to a far lesser extent, barometric pressure, are required to correct both the altimeter and the air speed indicator, but Wind Vector is most critical, since it is the primary force acting to deflect the course of an aircraft in flight. Despite the availability of instruments for navigation in flight, all flights begin with a dead-reckoned flight plan, in which the three Wind Triangle vectors are precisely plotted or computed.

Referring to the Wind Triangle inset on C-3, above: for a dead-reckoned flight along a desired Course Vector of 272.5 degrees True (272.5 degrees clockwise from True North) at 110.6kt (knots or *nautical* miles per hour) ground speed from Apra harbor to Cavite, *Hawaii Clipper* would have planned to fly along the true Heading Vector, which would compensate for wind drift produced by the Wind Vector. To put it simply, the navigator would have plotted a Heading Vector of 269 degrees True, at 128kt True Air Speed, such that the average Wind Vector of 247 degrees True at 19kt would have blown the Clipper back on the nominal (and desired) Course Vector, even as the flying boat attempted to fly the Heading Vector. Wind Vector values, as used for all the calculations within this analysis, are based upon winds aloft logged at Guam and Cavite immediately prior to the flight, and upon winds reported by *Hawaii Clipper* just before radio contact was lost. It should be noted that *wind direction*, unlike sea drift, is logged as the direction *from which it blows*. The wind is blowing *toward* 67 degrees True (67T), but is always logged, by convention, as being *from* the reciprocal (i.e., the sum of 67T + 180), or—247T.

Because dead-reckoning is precise, at least on paper, analysis of a dead-reckoned flight is easier than analysis of flights involving in-flight corrections based upon celestial or landmark sightings, radio bearings and the like. However, for purposes of analysis, the most important flight parameter, True Air Speed (TAS), the speed at which the aircraft actually moves through the air, is generally not reported and must be calculated from other parameters. In fact, *without* the TAS, this analysis would have been impossible. However, data from the PAA's *Engineering Report* and the ASB's *Preliminary Report*, as well as the data received from *Hawaii Clipper* (all noted in the Appendix), make it possible to calculate the actual TAS easily, and with a high degree of reliable accuracy:

Since fuel economy was (and always is) an important factor in flight integrity, the first consideration for the Clipper's flight plan was the selection of the TAS which would provide the longest cruising range for the fuel aboard at the planned cruising altitude. For a Martin *M-130* Clipper, the best range could be attained at a *density altitude* of 7800 ft, which was the optimum effective altitude for supercharger efficiency. Based on 2550 gallons of gasoline reported aboard, *Hawaii Clipper* had a maximum range, at a density altitude of 7800ft (actually 8700ft true altitude, as corrected with a reported air temperature of 13 deg. C, and the average barometric pressure) of 2258 nautical miles in 17.5hr. Simple math (2258nm / 17.5hr) yielded a TAS of 129.03kt, but, as *Hawaii Clipper* reported her altitude as being at between 9100ft and 10,000ft altitude during the last flight, the optimum TAS would have been somewhat lower than 129kt, owing to the reduced supercharger efficiency above 8700ft true altitude (which was equivalent to the calculated density altitude of 7800ft).

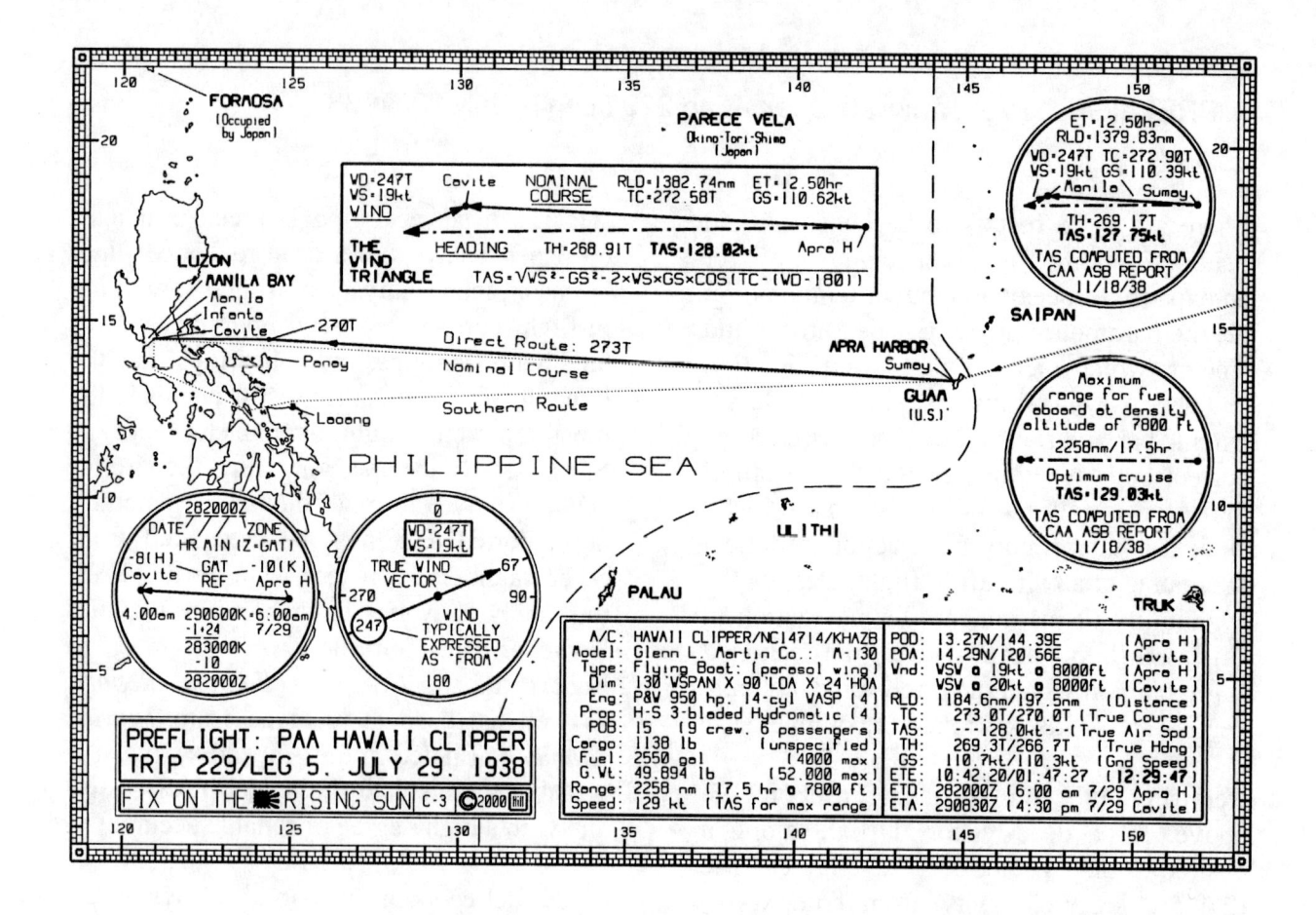

For the purposes of this analysis, calculating the optimum TAS for a given altitude can yield only a theoretical value, whereas the actual TAS of the Clipper in flight must be derived from actual reported flight data. The math becomes much easier in this respect, and requires no complicated corrections.

Computation of the Clipper's TAS for Trip 229 (from *projected performance data*, as provided in the 1938 report) indicates a TAS of 127.75kt, as shown in the top right inset. Computing the TAS for the *nominal course*, direct from Apra Harbor to Cavite, during a 12.5hr flight, yields a TAS of 128.02kt, as shown in the Wind Triangle inset. Adopting a nominal TAS of 128kt, the Wind Triangle was recalculated to provide the data in the large data block at the bottom of the chart. This final set of computed data reflects a division of the leg into two sub-legs, one from Apra Harbor to a point north of Panay Radio, along course 273T, and a final sub-leg due west, at 270T, to Cavite. It should be noted that, due to correction for temperature and barometric pressure, a TAS of *128kt* at 9100ft, is a bit higher than the Indicated Air Speed (IAS) of about *122.4kt*, which would have been seen and maintained on the Air Speed Indicator *and which is remembered by former PAA pilots*. As will be seen, the TAS of 128kt, as calculated from projected data, has been positively confirmed by data received from the Clipper, in her last report.

Additionally, this analysis uses the same convention for time-keeping as that used by PAA: Greenwich Mean Time (GMT), which is indicated by the suffix "Z," as shown in the Time Reference inset, above. This inset provides general information relating to the application of GMT within this analysis. PAA's use of GMT eliminated the confusion that tends to occur when multiple time zones are referenced—as did occur during the confusing Earhart search in July, 1937.

8. LAST LEG TO OBLIVION: Final Hours of the *Hawaii Clipper*

The *Hawaii Clipper* departed from Sumay, Guam at 281939Z (5:39am, July 29), but did not actually take off from Apra Harbor until 29 minutes later, at 282008Z (6:08am). The delay seems excessive, but may have been due to ship traffic in Apra Harbor. In any case, it appears that a 6:00am take-off was intended, but was delayed for some reason. A story later circulated, claiming that the take-off run had taken perhaps twenty-five percent longer than usual. This indicates an unbalanced, tail-heavy condition, caused by an *unexpected* excessive weight in the tail.

Communication was maintained on the hour and the half-hour, with RDF bearings being taken between the regular reports. Radio Sumay provided both of these services until 0200Z, when the Clipper reached mid-flight between Sumay and tiny Panay Island near Luzon, in the Philippines. Radio Sumay then stood down, as Radio Panay and the Makati RDF began providing communications and RDF services, respectively. Andre Priester's PAA *Engineering Report* of August 2, 1938, includes none of the early positions received by Radio Sumay (which could not receive the Clipper after 0200Z), noting only that, "all positions since departure from Guam were computed by dead reckoning with the exception of one sun line." The report cites the last five positions (sent to Panay) as:

Pos. A: 0200Z: Lat. 12.20N Lon. 134.26E
"Weather - squalls all around"
Pos. B: 0230Z: Lat. 12.00N Lon. 133.36E
Pos. C: 0300Z: Lat. 12.05N Lon. 133.28E
"Clouds 10/10ths above 10/10ths below
...10,000 feet [altitude]"
Pos. D: 0330Z: Lat. 12.15N Lon. 131.37E
Pos. E: 0400Z: Lat. 12.27N Lon. 130.40E
"10/10ths...stratocumulus [above].
10/10ths cumulus [below].
Altitude 9100. Rain."

Priester notes in his PAA Report that the last position was "on the regular southern course between Guam and Manila." This was quite an unusual assessment, because, while it is true that Pos. D and Pos. E lie quite near the obsolete Southern Route, they can hardly be said to be representative of that route, which provided access only to the San Bernardino Strait. Had the Clipper actually been on the "regular southern course," as Priester put it, communications would have been provided through Laoang, rather than through Panay.

Curiously, the November 18, 1938 report of the CAA's Air Safety Board (similar to the National Transportation Safety Board) cited only the *final* position, Pos. E (0400Z), which had already been published, but it did cite all of the details of the 0400Z report:

"Flying in rough air at 9100 feet. Temperature 13 degrees Centigrade. Wind 19 knots...from 247 degrees. Position Latitude 12.27N, Longitude 130.40E *dead reckoning*. Ground speed made good 112 knots. Desired track 282T. Rain. During past hour cloud conditions have varied. 10/10ths of sky above covered by strato cumulus clouds, base 9200 ft. Clouds below, 10/10ths of sky covered by cumulus clouds whose tops were 9200 feet. 5/10ths of the hour on instruments. Last direction finder bearing from Manila 101 True [author's italics]."

That "Last direction finder bearing from Manila" was taken from Makati at about 0345Z. Although the bearings from Makati to Pos. D and Pos. E are 101.36T and 101.52T, respectively, an RDF bearing of 101T was close enough to "confirm" the Clipper's reported position and course. The ASB Report also noted that at 0330Z, at Pos. D there were "1420 gallons of fuel aboard, sufficient for 10.1 hours of normal cruising."

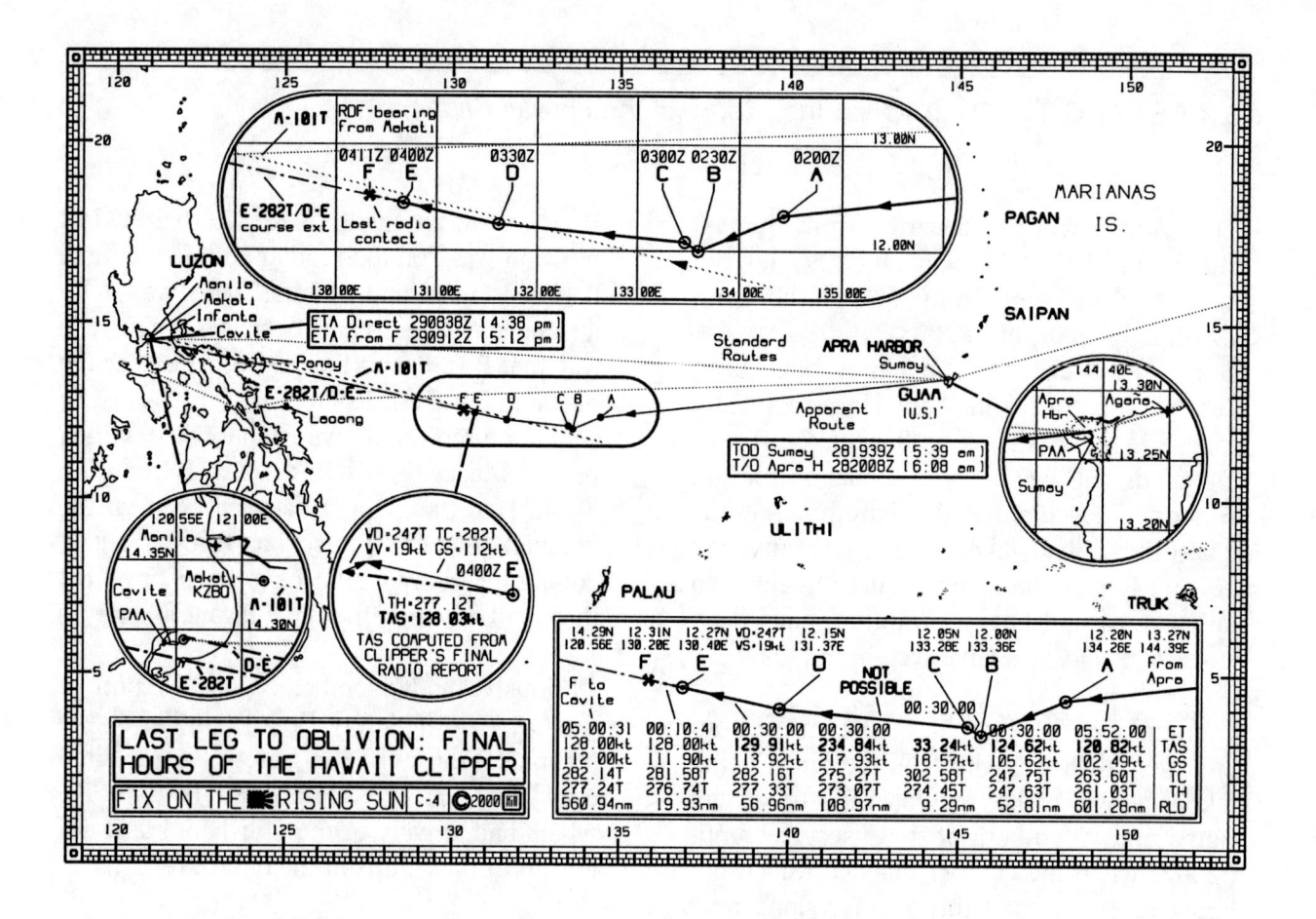

It also cites the Clipper's last message, received by Fernandez, at Panay, at 0411Z:

"Stand by for one minute before sending as I am having trouble with rain static."

Extending a course from Pos. E at 282T, the line passes 2 miles south of Cavite and can be terminated within a thousand *feet* of the crossing of Latitude 14.30N, Longitude 120.45E: a convenient point on any chart. However, the extension from D to E passes so precisely through the Cavite landing area that D-E EXT, not E-282T, may have been the intended course reference.

The Clipper's 128kt TAS is confirmed using data received in the 0400Z report, as shown in the inset. At 128kt TAS a course from Pos. E can be extended to 0411Z, yielding the position, as radio signals ended, as:

Pos. F: 0411Z: Lat. 12.31N Lon. 130.20E.

With regard to the flight analysis, the flight from Guam to Pos. A was apparently made at a TAS of just under 121kt, which seems low, unless the flight was not direct from Guam. From A to B, a TAS of 124.62kt also seems low, *but the legs from B to C and C to D are absolutely in doubt*: B to C might have been the result of flying in circles, as if avoiding squalls, *but C to D is simply not possible*, requiring a TAS of 235kt, which was well beyond the Martin *M-130*'s 160kt top speed. The ground speed From D to E is also doubtful, since the calculated ground speed of 113.9kt, reported at Pos. D at 0330Z, is 2kt higher than the 112kt "speed-made-good" reported at Pos. E at 0400Z.

As to the unidentified "sun line" position, which Priester mentioned in his report, it may well have been the usual maritime noon position at Pos. A (0200Z), but, considering the reportedly inclement weather, any sun-line observation seems quite unlikely.

9. MANILA BAY: D-E EXT and E-282T Courses to Manila Bay

Although the former Pan American Airways Captain, Harry Canaday, asserted, in 1992, that all Clipper flights to Manila Bay used the city of Manila as the navigation arrival point, the final course of *Hawaii Clipper*, as reported from the Clipper at 290400Z, was clearly directed to the Cavite area, with the 282T course from Pos. E being directed, apparently, to the crossing of Lat. 14.30N and Lon. 120.45E, in the center of Manila Bay—missing by only about 1000 feet, as if a pencil line had been drawn on a chart, from Pos. E to that crossing.

It is of further interest that the extension of a course line extended from D to E (the last two reported positions and the only ones determining a course to the Manila area) passes precisely through the landing area just outside of Canacao Bay, at Cavite. If Captain Canaday's contention was correct, then the course should have passed through Manila, rather than Cavite. If his contention was not correct for every flight, then it appears that the D to E line represents a true course, not merely to the PAA terminal (as a fixed landmark), but to the *landing area* just outside Canacao Bay—a bay area having no fixed landmark for navigational purposes.

The use of a flying boat *landing area* in a body of water, as an indicated arrival point for navigation, is a curious factor of the flight report, one which is quite important, when the rest of the reported positions are considered—as will be shown. Still more important is the fact that the Clipper was clearly not indicating a course along the old Southern Route, as Priester's *Engineering Report* indicates. Nor was the Clipper even following the Direct Route, but, rather, was following a route that would take it across the most hazardous mountain terrain in all of southern Luzon. And it may be assumed that Panay had already advised the Clipper of the weather conditions during previous contacts; that is, the skies overcast, with a 2500 foot ceiling and light rain. It was most unlikely that Captain Terletsky, a pilot with 9200 hours of flying (1626 hours on the Pacific Clippers, alone), would risk ending his long career—and, perhaps, his life—by dropping down through the overcast into mountainous terrain—especially when his radio operator was experiencing sufficient rain static as to seriously impair radio communications (as was, in fact, reported at 0411Z, following the Clipper's final flight report).

As to the timing of the final report, some note should be made of the radio reporting process. From the PAA *Engineering Report*, prepared on or before August 2 (five days after the Clipper was lost), it is known that the 0400Z report was filed at 0403Z. That is, the Second Officer (and also the Navigator, by maritime convention) had completed his navigational plot after all data was acquired at 0400Z, added the data to the message blank and immediately passed it down to the Flight Radio Officer (FRO) who began *sending* the report out at 0403Z.

It should be noted that the FRO was seated directly behind the co-pilot's seat on the right side of the "bridge," next to the rear window. The Navigator's chart table was located on the cabane deck, in line with the propellers, just behind and slightly above the FRO's position, but well forward of the Engineer Officer's position, which was just under and slightly forward of the radio mast.

It is significant, for this analysis, that the Navigator and the FRO were stationed in close proximity to each other and, therefore, quite able to communicate closely with each other in the interest of moving radio traffic.

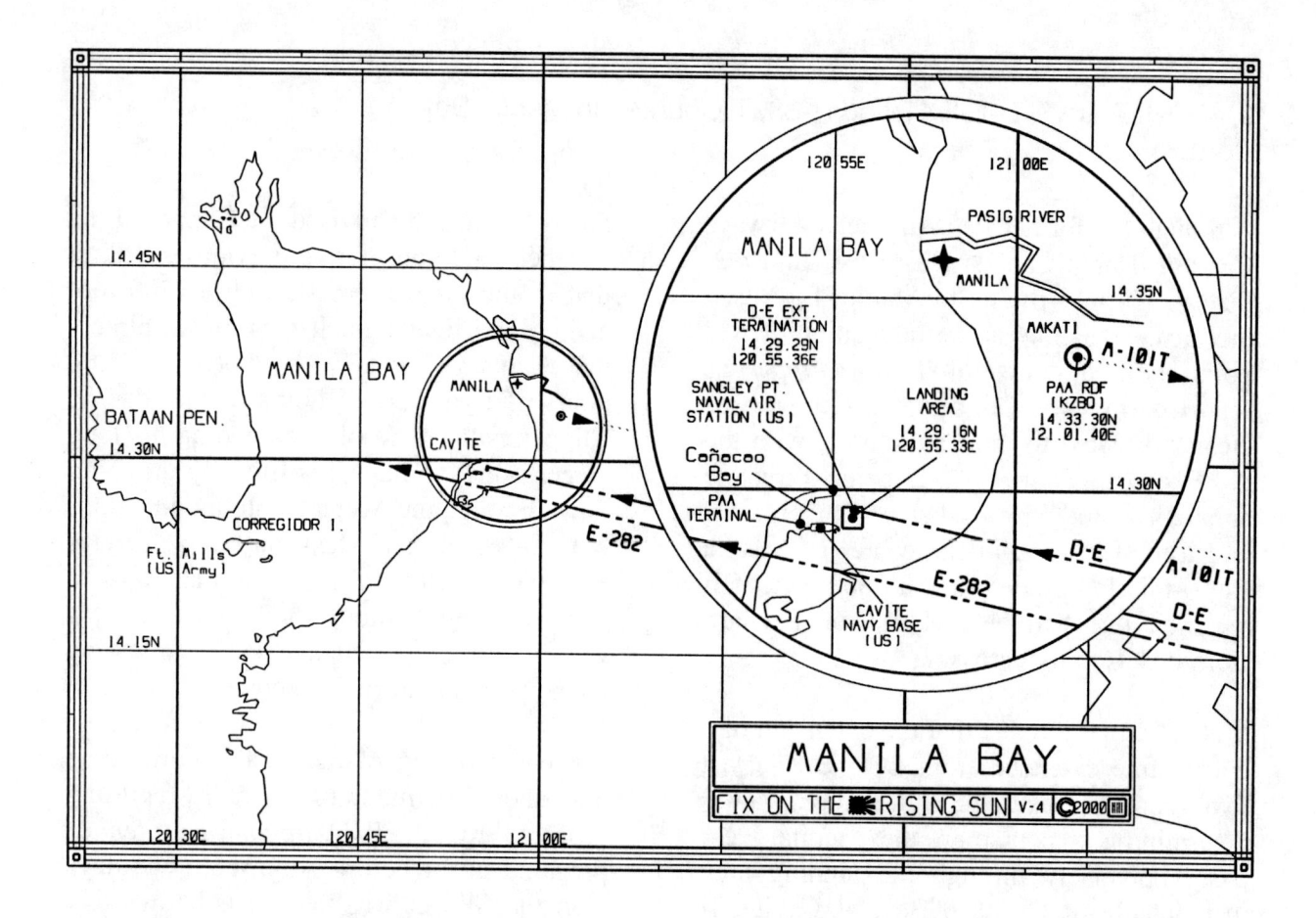

With regard to the Clipper's last message, it took eight minutes (at the usual rate of fifteen words per minute) for the FRO to establish contact with Radio Panay and to transmit the message, which was completed, as the August 2 report indicates, at 0411Z.

During that same minute, at 0411Z, Panay contacted the Clipper to send the weather report, but the FRO asked Panay to stand by, reporting that the Clipper was experiencing rain static (the result of fast moving air and rain over a receiving antenna). Panay called again, at 0412Z, but never received a reply. Hence, the time of the disappearance was logged as 0411Z, the very minute during which the last of *Hawaii Clipper*'s radio transmissions were received by any station.

As to the convention used to express courses for this analysis, three simple expressions have been adopted: "D-E" represents a line between two lettered positions, the direction of the course being from the first lettered position to the second. The expression "D-E EXT" represents a line extended from the second lettered position along the course defined by the relationship of the first to the second. The expression "E-282T" represents a course from the lettered position along a line defined by a True compass bearing, such as 282T, which may be read as "282 degrees True," and which sets 282 degrees clockwise from True north. This expression is also used with regard to the radio bearing reported from Makati, which is expressed as a Great Circle "course" from the Makati RDF (M) and radiating *initially* on a True compass bearing (101T), as above. It should be noted that, although the radio signals traveled from the Clipper to the Makati RDF, such radio bearings are expressed, by convention, as Great Circle bearings from the RDF toward the source of the signal. Such bearings generally appear as *curved* lines on Mercator navigation charts.

10. AFTERMATH: A Divided Search and Flight Reports in Doubt

In the aftermath of *Hawaii Clipper*'s loss, two traces were soon reported, which served only to divide the search. First, a telephone company employee on Lahuy had reported hearing a "large airplane" above the overcast at "about" 3:00pm Manila time (290700Z). This report was given much credence during the search—and even later, when the ASB Report was released—so it may be assumed that both the report and the witness had been determined to be reliable. For those readers who do not remember the sound of four-engine aircraft, it will have to be taken on faith that the sounds of single-, two- and four-engine aircraft are all quite distinctive. Further, although many Filipinos had come to know the Clippers and the sound of their engines quite well during the previous three years, four-engine aircraft were not at all commonplace in 1938. A review of military, commercial and civil flights on that day had decidedly ruled out the presence of other aircraft near Lahuy Island at about 0700Z. Curiously, however, if the big aircraft, heard above the overcast, was, indeed, the *Hawaii Clipper*, she was about 39 minutes ahead of schedule (from Pos. E) and flying at a TAS of about *154kt*, very nearly the top speed of the aircraft. Still, the report was considered to be of sufficient importance that the U.S. Army Air Corps initiated a widespread *land search* over some 175 nautical miles of the mountainous Luzon jungle, along the route from Lahuy to Cavite.

Second, the U.S. Army Transport, *Meigs*, which, was somehow known to be near *Hawaii Clipper*'s regular flight path, was directed to proceed to the last reported position, Pos. E and to commence a sea search. According to Priester's PAA Report, "at 0430 GCT [0430Z] the steamer U. S. Meigs en route to Guam was immediately called...[and]...immediately proceeded to the site of the last position..." The ASB Report stated later, however, that "Meigs altered her course at 0740 G.C.T. [some three hours later]...and proceeded to the Clipper's last reported position..." About 29 hours after the Clipper's last transmission, at 300910Z (5:10pm the next day), the *Meigs* discovered an oil slick, 500 to 1500 feet in diameter, at or about Lat. 12.11N, Lon. 130.33E. From the reported current set of 1kt at 140T, the slick can be (and most certainly was, then) calculated to have been positioned, 29 hours earlier, at Lat. 12.33N, Lon. 130.14E, along the Clipper's reported course and about 6 nautical miles (3 minutes flight time) from the calculated 0411Z position at Pos. F.

Meigs' officers took a small boat into the slick, collected oil samples, and noted an odor of *gasoline*, as well. Although there were absolutely no human remains or debris *whatsoever* in the slick, the slick's origin estimate, so near the Clipper's final position estimate, provided compelling evidence to justify a *sea search*. The U.S. Navy had already dispatched some fourteen surface vessels and submarines of the Asiatic Fleet to conduct a search, and this was maintained in various sea areas for about a week, ultimately extending far to the north of the last position. The search was, thus, divided, and no further traces of the *Hawaii Clipper* would *ever* be officially reported.

As to a continued analysis of the flight, itself, the leg from Apra Harbor to Pos. A was dead-reckoned at a TAS of 128kt and should have required only about 5hr 29mn at a ground speed (GS) of 109.7kt, rather than 5hr 52mn at 102.49kt, as was shown on Chart C-4. For some reason, the Clipper appeared to have killed about 23 minutes en route to Pos. A. But neither of these figures squares with newspaper reports that cited a position report sent to Sumay at 282100Z, about 52 minutes after take-off from Apra Harbor. This report noted that the Clipper was at a cruising altitude of 10,000ft, but was operating at only 105kt ground speed, or about 123kt TAS, some 5kt slower than desired, and at about the same speed as that already calculated between the later reported positions from A to B at 0200Z and 0230Z.

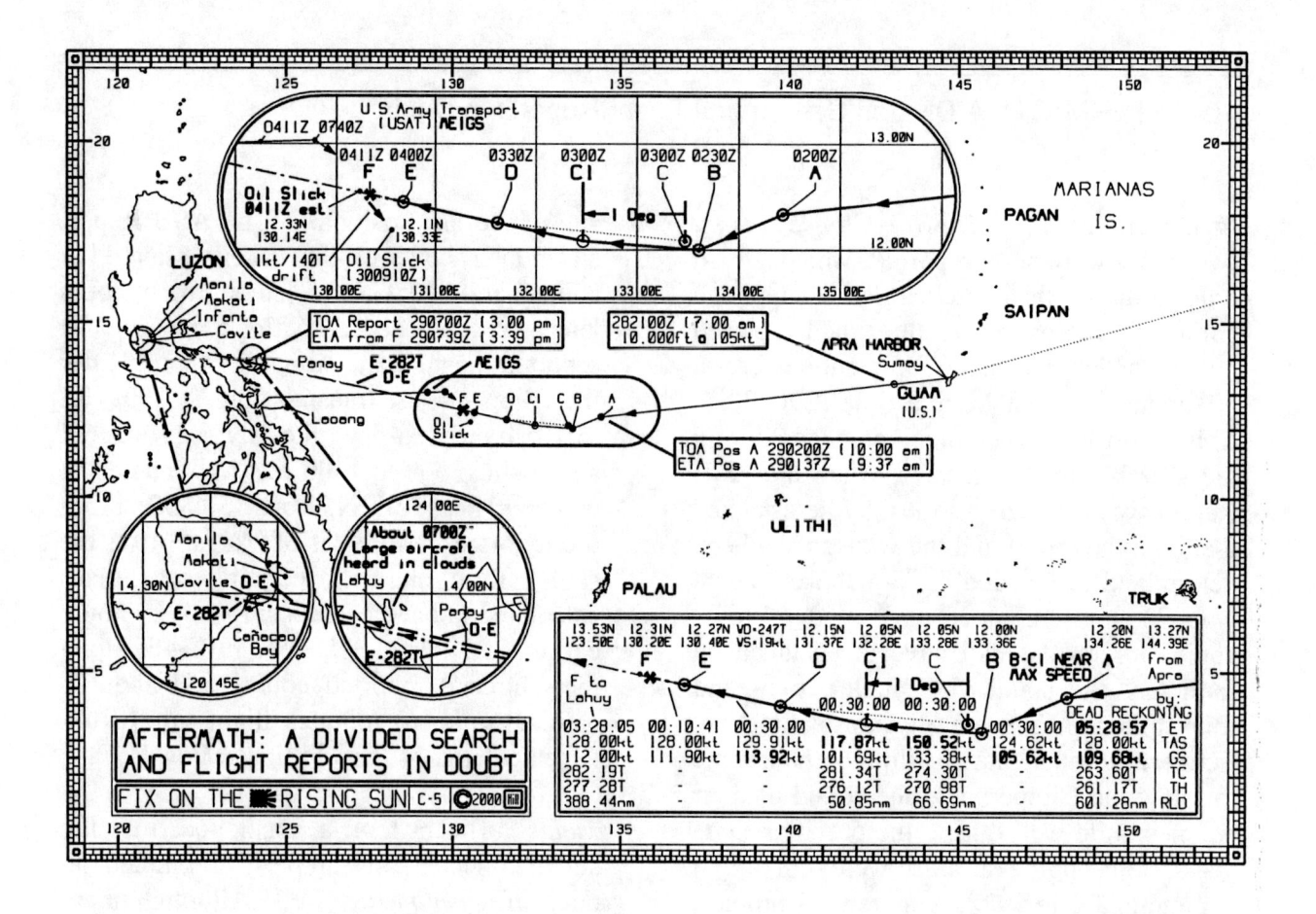

Hawaii Clipper, inexplicably, had just been "loafing" along, on reaching 10,000ft, and, although PAA Captains were never required to operate the Clippers on a tight schedule, it seems unusual, given the rather benign pre-flight weather reports, that Captain Terletsky would have elected to add even more time to a twelve and a half hour flight.

In an attempt to resolve the flight analysis problem arising out of the reported positions of Pos. B and Pos. C (only 9.29 nautical miles apart), it seemed possible that Pos. C might have been reported as being 1 degree of longitude (1 digit in the 0300Z report) to the east of the Clipper's 0300Z position. Consequently, the route from A to E was re-plotted, with Pos. C1 replacing Pos. C, 1 degree to the west. This made the C1 to D leg at least possible, but the B to C1 leg then required a TAS of very nearly the top speed of the Clipper—possible, but still too high. [*This was not an attempt to "edit" any of the evidence, but strictly to determine whether such an error was mathematically feasible.*]

Keeping in mind that the route from Guam to Pos. E was one of dead-reckoning (that is, a mathematically precise model of a route), and that the most desirable TAS was 128kt, it was quite disturbing to find that none of the various legs, including that from D to E, showed any firm mathematical relationship, either to each other or to the most desirable True Air Speed. The flight reports, *closely examined*, appeared to be in serious doubt.

11. 3 CLIPPER APPROACHES TO CAVITE: 1 Secure, 1 Efficient—1 Risky

The report of a Lahuy telephone company employee, regarding a "large airplane" heard above the overcast skies at "about 3:00pm," (Manila Time, or, 0700Z), coupled with the investigation which ruled out other known aircraft in the area, established two facts:

First, that the aircraft that was heard was either the *Hawaii Clipper* or an aircraft not of U.S. or Philippine origin, and second, that southeastern Luzon was "socked in" under heavy overcast, a fact which was confirmed by PAA records (and as noted in the ASB Report). Only twelve hours earlier, Cavite had been reported to have had light (1/10) cloud cover with a ceiling (clear space under the clouds) of 2500 feet, but conditions can change rapidly in the Philippines, which was one of the reasons that weather reporting and reliable radio communications were so important to Clipper operations.

The Lahuy report makes it clear that the reported approach of *Hawaii Clipper* (if this was, in actual fact, the *Hawaii Clipper*) was especially imprudent, considering that every major mountain top on the route was within the clouds—making a reduction in altitude (for final approach) hazardous, at best.

We are expected to believe that the entire 200 mile approach, from first landfall, at Catanduanes Island, to the landing area at Cavite, was to be made over some of the densest mountain jungles in the world, and above a heavy overcast which afforded no ceiling between the major mountain tops and the rain clouds (which extended up to 9200 feet altitude and beyond). With rain static having interfered with communications only hours before and (based upon the time of the Lahuy report) flying, inexplicably, at very nearly the absolute top speed of the aircraft, such a risky approach appears most unlikely.

In view of the fact that PAA had effectively obsoleted its secure Southern Route through the San Bernardino Strait only after greatly improving the safety of its efficient Direct Route (with the addition of stations at Panay and Infanta), it seems quite improbable that Captain Leo Terletsky, a reportedly cautious Russian immigrant with a long history of service with PAA, would risk his aircraft, his crew, his passengers, and his own life, to make a blind, direct run at Cavite over such dangerous ground. This was, after all, a flying boat, whose greatest claim to safety was its capacity to make a safe landing on any sizable body of water.

All pilots of land-based aircraft instinctively scan the ground beneath them during flight, noting possible landing areas in the event of a forced landing. This is true for pilots of smaller aircraft flying over rugged terrain, even today, but it was seldom a concern for Clipper pilots, so long as they were flying with an expanse of water under the keel. This was noted even by PAA officials, not long after the search for the *Hawaii Clipper* had ended, illustrating their own doubts of the likelihood of such a dangerous course, as was reported during the Clipper's last flight.

It seems apparent that the "large aircraft" heard over Lahuy was probably not *Hawaii Clipper*, especially as, during the sixty years since her disappearance, no trace has ever been found near Luzon. The only remaining probability is that the aircraft was neither U.S., nor Philippine, nor from *any* "friendly" nation (which would have acknowledged an over-flight and so prevented the useless air search of southeastern Luzon). The aircraft heard over Lahuy, therefore, if it were not the *Hawaii Clipper*, could have been none other than an *unfriendly* aircraft—as will be demonstrated in subsequent charts.

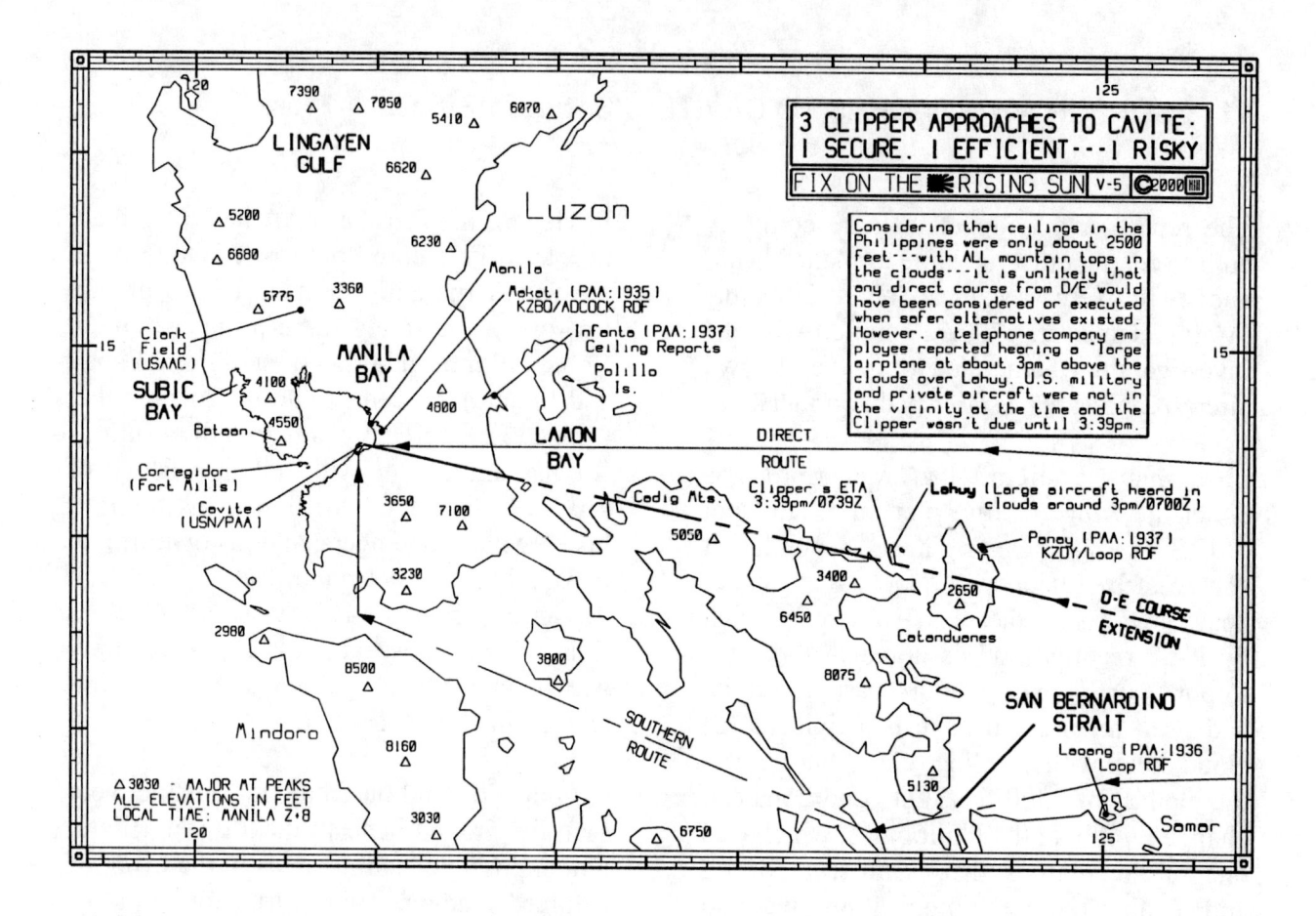

It may be difficult in these days of ground- and on-board radar to understand the great difficulties—and dangers—encountered by flyers during the 1930's, navigating through inclement weather and over rugged terrain. Even during the later Pacific War, when radar was developed for use in aviation, the dangers still remained, and, in January of 1942, in a heavy overcast, the *Philippine Clipper* lost her way, without knowing it, en route to Alameda. Denied permission to proceed to China Lake, PAA's alternate landing area, she was cruising in what she thought was a safe pattern, waiting for the weather to clear below her in San Francisco Bay—when she disappeared. It required two full weeks for searchers to locate her—on a mountain, *far to the north of the Bay Area.*

That the problem was indeed serious in 1938 is illustrated by a story carried in the *New York Times* on July 31, 1938—just three days after the *Hawaii Clipper* disappeared:

"Within thirty days, transport airplanes winging down through Newhall Pass into the Union Air Terminal [in Los Angeles] will glide along an invisible radio track just like toboggans along a slide. The 'blind landing' device, perfected by United Airlines and the Bendix Corporation, is being installed at Burbank...[and will]...be ready for tests late in August. Radio markers along the airway approach to the field, and a constant wireless beam emanating from the station will guide transports down through tricky weather, thus abolishing the present procedure of lowering a ship *through the murk* [author's italics]."

This use of the phrase "through the murk" echoes the title of Jordanoff's 1938 textbook on aviation weather, *Through the Overcast,* and, if this mortal hazard was a matter of great concern in southern California, it was, surely, a matter of far greater concern over the vast wilderness of southeastern Luzon.

12. THE CHANGING FACE OF TRUTH: Analysis Raises More Doubts

Many details of the Clipper's last flight have failed to survive analysis—past and present: a 1938 laboratory analysis of oil taken from the slick found by the *Meigs*, for instance, revealed that the oil had not originated with the Clipper. And empirical tests on a similar slick created in San Francisco Bay (with a mixture of oil and gasoline) indicated that the gasoline evaporated within five minutes, leaving no lingering odor, as the *Meigs*' officers had noted previously. Therefore, the CAA's Air Safety Board, apparently taking into account the lack of debris or human remains, discounted the slick entirely (but without doubting its existence), and retained only the Lahuy report as reliable evidence.

As a footnote to the gas/oil slick, it should be noted that the *Meigs*' observations and the San Francisco experiment suggest that the original slick may not have been a static surface slick, but a dynamic slick, created by oil and gasoline supplied from some limited source beneath the surface. The complete disappearance of the slick, two days after the *Hawaii Clipper*'s loss, indicates that the oil found in the slick was volatile and of a light grade, which dissipated nearly as fast as the gasoline, and that both the oil and gasoline sources were depleted at the same time.

The 2 knot discrepancy in the ground speed, from D to E, as already noted in Chart C-4, has succumbed to more recent analysis and places Pos. E in question as a true position achieved at 0400Z. The problem, in this case, turns on the lower resolution of air navigation, which expresses all positions in terms of degrees and minutes, rather than in degrees, minutes and *seconds*, as in marine navigation. Minutes of arc yield an accuracy only to plus or minus 1/2 of a nautical mile, whereas seconds yield far greater accuracy, theoretically to within about 50 feet.

The right-hand circular chart inset shows a close-up of Pos. E with an hypothetical Pos. E1, 1 minute to the east. Four courses are also shown: D-E1 EXT, D-E EXT, D-Cavite Landing Area and D-282T. At 128kt TAS, any course from Pos. D would reach the oblique dashed line in exactly 30 minutes, at the reported ground speed of nearly 112kt.

Considering the ideal D-Cavite course, it should be clear that Pos. E1 represents Distance Made Good along that course more closely than Pos. E, and the lower data block shows that the D-E1 leg reflects a TAS of 127.98kt—almost precisely 128kt. However, as the left inset shows, D-E1 EXT misses Cavite by a large margin, while D-E EXT is a perfect match for a D-Cavite course.

It is clear that, in reporting Pos. E, the Clipper's navigator was expressing Course Made Good rather than the Distance Made Good, which would have been indicated by Pos. E1. But Distance Made Good is the critical factor in dead-reckoned, single-RDF navigation, not only in Noonan's method, but in all cases, because distance (although usually reliable due to fixed engine settings) cannot be confirmed during the flight if the monitor RDF lies close to the course or the destination, as with Makati. The navigator seems to have been more intent on showing a D-E pointer to Cavite, than upon reliably reporting his Distance Made Good, as would have been the case, had he specified Pos. E1 for the 0400Z position. Pos. E appears to be a *reconfiguration* of the appropriate Pos. E1.

As to Pos. B and Pos. C, analysis now places them even further in doubt, especially in light of the discovery of the long-sought math match between the flight data (as reported in 1938) and the analytical dead-reckoning data (as recently calculated).

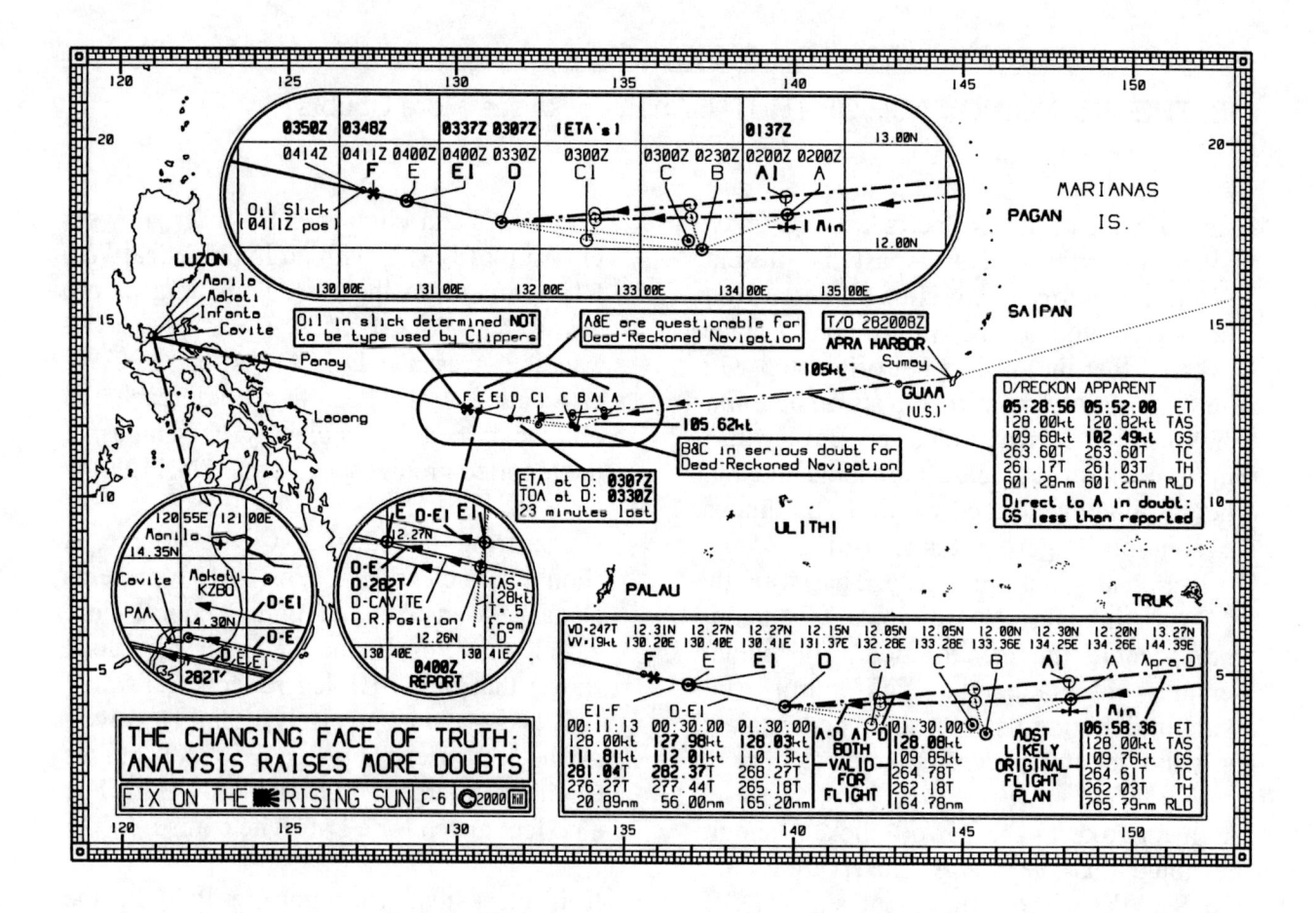

THE MATH MATCH: An examination of the Direct Route from A to D reveals that *only this Direct Course* could have been flown in the time allotted (1.5hr with winds as noted) at 128.03kt TAS—a near-perfect match to the nominal 128kt TAS! Flying other than A-D direct (A-B-C-D, e.g.) would always require an increased TAS, indicating that Pos. D may not have been a *resultant* route point, but a very *specific* intermediate destination which was to be reached on the hour or half-hour—*so as to be reported.*

If the flight plan had initially postulated the Direct Route from Apra Harbor to Pos. D (e.g., to avoid weather) then Pos. A would have occurred at Pos. A1, which is only 1 minute of arc west, and 10 minutes north, of Pos. A, but still so situated as to conform to a TAS of 128.08kt, from A to D. And, even though a shortfall of a minute of estimated flight time, 5hr 28mn 56sc to Pos. A plus 1hr 30mn to Pos. D) seems excessive for a dead-reckoned course, the route would have yielded a near-precise 7hr (6hr 58mn 56sc) flight from Apra to D.

However, such a flight would have required a GS (ground speed) of over 109kt, which is at odds with the reported initial GS of 105kt and the calculated 105.62kt GS from A to B. Yet the actual average GS between Apra and Pos. A was only 102.49kt, which is at odds with the early 105kt report and *puzzling*, in that the *Hawaii Clipper* took 23 minutes longer than necessary to reach Pos. A and, altogether, 7hr, 22mn to reach Pos. D.

NOTE WELL: The early report of a 105kt GS (about 123kt TAS) seems to indicate that *Hawaii Clipper* was cruising *below* a 128kt TAS. More important, still, it indicates that she was cruising *faster* than the 102.49kt GS that was calculated for Apra-A/A1 (above). This indicates (as will be seen on C-12) that the Apra-A/A1 route *was not direct!*

13. THE NAVIGATOR'S "SUN-LINE": A Graphic "Rising Sun"-Line

With so many facts in doubt, with regard to the *Hawaii Clipper*'s last flight, the reported "sun-line," as noted in Andre Priester's PAA *Engineering Report* of August 2, certainly demands attention, despite the unlikelihood of such a solar sighting ever having taken place (given the weather) and also despite Priester failure to note just when the sighting took place. In his report, he noted only that, "All positions since departure from Guam were computed by dead reckoning with the exception of one sun line."

Noon sightings of the sun are a maritime tradition, as well as an accurate and reliable navigation practice, but Pos. A or Pos. A1, which was reported as having been reached at 0200Z, appeared to be the logical place to begin an analytical search, even though it was reached at noon *Guam time*—not solar time. It was, after all, the position at which Sumay stood down and Radio Panay and the Makati RDF assumed responsibility for all flight communications in support of a Cavite approach—a critical point of transition.

Moreover, because the weather would seem to have precluded any sighting of the sun at that time, it appeared more likely that this "sun-line" might only have been Priester's subtle reference to the "Rising Sun,"—the Japanese metaphor for the Empire of Japan.

In his excellent (and highly recommended) 1980 book, *China Clipper*, Ronald Jackson suggested that *Hawaii Clipper* had been hi-jacked to Palau, in part, because Palau was the closest of the Japanese Mandates, but, more important, because Palau lay due south of Pos. A. Jackson had also associated the hi-jacking with the Clipper's noon position (Guam time), and so Palau appeared to be as likely a place as any to begin the search for Priester's elusive "sun-line."

Pos. A and Pos. A1 both represent precise intermediate positions, with respect to Pos. D, for a nearly exact 7-hour dead-reckoned direct route from Apra Harbor to the course-correction position at Pos. D. But a line due south from Pos. A1 crosses no particular landmarks or islands in Palau, except in the southern-most part of the group. However, a line due south from the actual reported location at Pos. A terminates precisely at the entrance to Palau's Malakal Harbor, where seaplanes could be, and often were, landed.

Initially, this seemed to explain why the Clipper had ultimately diverted to Pos. A, when it appeared that she had originally dead-reckoned an arrival at Pos. A1: That is, Pos. A1 could not provide the precision for an implied Line of Position due south, whereas Pos. A provides a pointer precisely to the harbor entrance. It may seem unusual to have used a body of water, with no fixed landmark, as a navigational point, but this also appeared to have been the case with the D-E LOP extension, which pointed to (or, passed through) the Cavite Landing Area.

Of course, a single Line of Position (LOP) cannot, in itself, determine a fix for position, and it could be argued that the relationship between Pos. A and the entrance to Malakal Harbor is little more than a coincidence. However, the disparity between Pos. A and the more appropriate dead-reckoned position at Pos. A1 indicates that *Hawaii Clipper* had some reason to abandon the direct course to Pos. D and to divert to Pos. A, or, simply, to report Pos. A, instead of Pos. A1.

In fact, as will be shown, *Hawaii Clipper*'s route was actually dead-reckoned *indirectly* from Guam and it was Pos. A, not Pos. A1, which became the *true* location for the Clipper's reported 0200Z position.

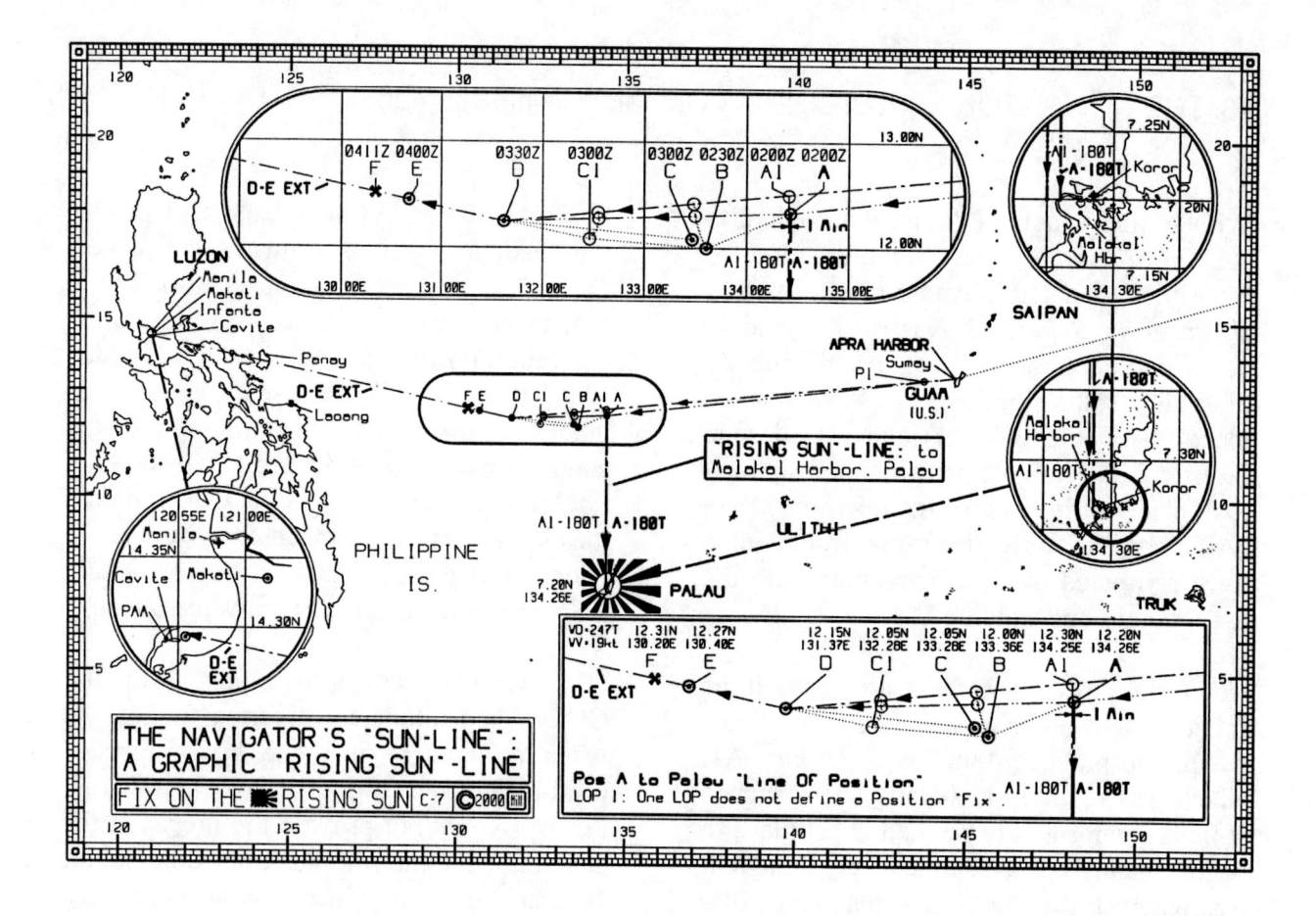

As to Jacksons's proposal that Koror may have been the destination for the hi-jacked Clipper, the Palau *Rising Sun*-Line, which Jackson must also have perceived, indicated, for him, a likely possibility. However, Koror offered only a haven—but certainly not a secure haven—for a hi-jacked Clipper:

In the early Mandates period, the Japanese put on a good show for the League of Nations, to demonstrate their responsible administration of the Pacific territories held in trust. Palau served, then, not only as the "front door" and the seat of civil governance of the Mandates (at Koror), but also as the showpiece of Japanese trust operations. By 1938, of course, Japan had withdrawn from the League and had no more need for dog-and-pony shows in Palau. Still, a certain openness persisted in the islands, whether for form or out of habit, even as Japan closed the curtains ever more tightly around other islands in the former Mandates.

For this reason, Palau would have been an unsuitable haven for *Hawaii Clipper*: there were simply too many outsiders passing in and out of Palau for a group of renegade naval aviators to risk landing a hi-jacked, but easily recognizable, Martin *M-130* and its crew. However, operations in support of a hi-jacking may have been another matter:

On August 5, 1938, the Associated Press released a Commerce Department report from the U.S. commercial attache in Tokyo, citing published Japanese reports that Japan intended to create an airlink connecting the Mandates with PAA's Clipper service out of Guam. Two routes were to be established to Tokyo from Palau: one route via Guam (the Clipper link) and one via Takao in Formosa. That Japan made this announcement to allay American suspicions after the hi-jacking is clear. Why the U.S. Government released it is not. But the route from Palau to Takao is of great interest, as will be seen further on.

14. PALAU: The "Rising Sun"-Line to Palau

Ronald Jackson's suggestion (in his 1980 work, *China Clipper*), that *Hawaii Clipper* might have been hi-jacked directly to Palau, hardly seems unreasonable, considering the close proximity of the Palau Islands to the Clipper's 0200Z position. The nature of his book, however, focused on a dramatization of the flight, interwoven with a history of PAA's Pacific venture, and, while he noted documented instances of Japanese sabotage, he avoided the technical analysis necessary for a discussion of complex hi-jacking plots. Therefore, Jackson overlooked the fact that, for a course south to Koror at 0200Z, the Makati RDF would certainly have detected the course change during subsequent radio bearings, taken between 0200Z and 0400Z.

Moreover Palau (Koror, specifically) was the "front door" to the Japanese Mandates, a center for trade, as well as for contact with the rest of the Mandates, themselves. It was at Koror, for example, in 1923, that the ashes of Major Earl ("Pete") Ellis, USMC, were released to a Chief Pharmacist's Mate of the U.S. Navy, for return to the U.S., via Tokyo. And, in the mid-1930's, the National Geographic Society was able to provide a strong pictorial story on the Mandates, by examining and photographing conditions within the entire Palau group. Palau was hardly wide open, by any means, but it did entertain sufficient foreign trade and traffic to make it inappropriate as the destination for a stolen aircraft, especially a flying boat as distinctive as *Hawaii Clipper*, one of only three Martin *M-130*'s, then the largest flying boats in service, anywhere in the world.

With regard to the "openness" at Palau, the Ellis case is somewhat representative, and certainly well known. Shortly after WWI, when Japan had convinced the League of Nations to *mandate*, to Japan, the protective sovereignty of German colonial territories in the Pacific, a number of forward-thinking Americans, Earl Ellis among them, began to realize that intelligence operations in the Mandates were necessary, in order to protect U.S. interests in the Pacific, in general, and the sea route to the Philippines, in particular. It is somewhat unclear whether Major Ellis' operation was mounted independently, or under official orders, but his actions have since been recognized by the Marine Corps.

Operating in various guises, Ellis toured the Japanese Mandates and, although the true extent of his travels in the Pacific may never be known, post-WWII reports have placed him in a number of locations throughout the Mandates. It is doubtful that he considered Palau to be of interest, from an intelligence standpoint, owing to its relative openness, but, in mid-1923, the U.S. naval attache in Tokyo was advised that Ellis had died of unknown causes at Palau. Some reports have since attributed his death to complications due to alcohol abuse, which may, for non-medical reasons, have led to his un-masking.

Japan granted permission for one American to travel to Koror by commercial steamer and to accompany Ellis' remains from there to Tokyo and thence, to the United States. The individual assigned to sail to Koror was Navy Chief Pharmacist's Mate Lawrence Zembsch, who was then stationed at the U.S. Navy hospital in Yokohama (established in WWI, when Japan was a wartime ally), and whose orders, based on his medical rating, called for an investigation into Ellis' death.

Advised, subsequently, of Zembsch's return to Japan, American authorities were quite disturbed to find Chief Zembsch standing motionless and alone on the dock, detached from reality and holding in his hands the box containing Ellis' ashes. A vacant expression seemed to indicate that he had been drugged, and, although his speech proved generally incomprehensible, it soon became apparent that he had no real memory of recent events.

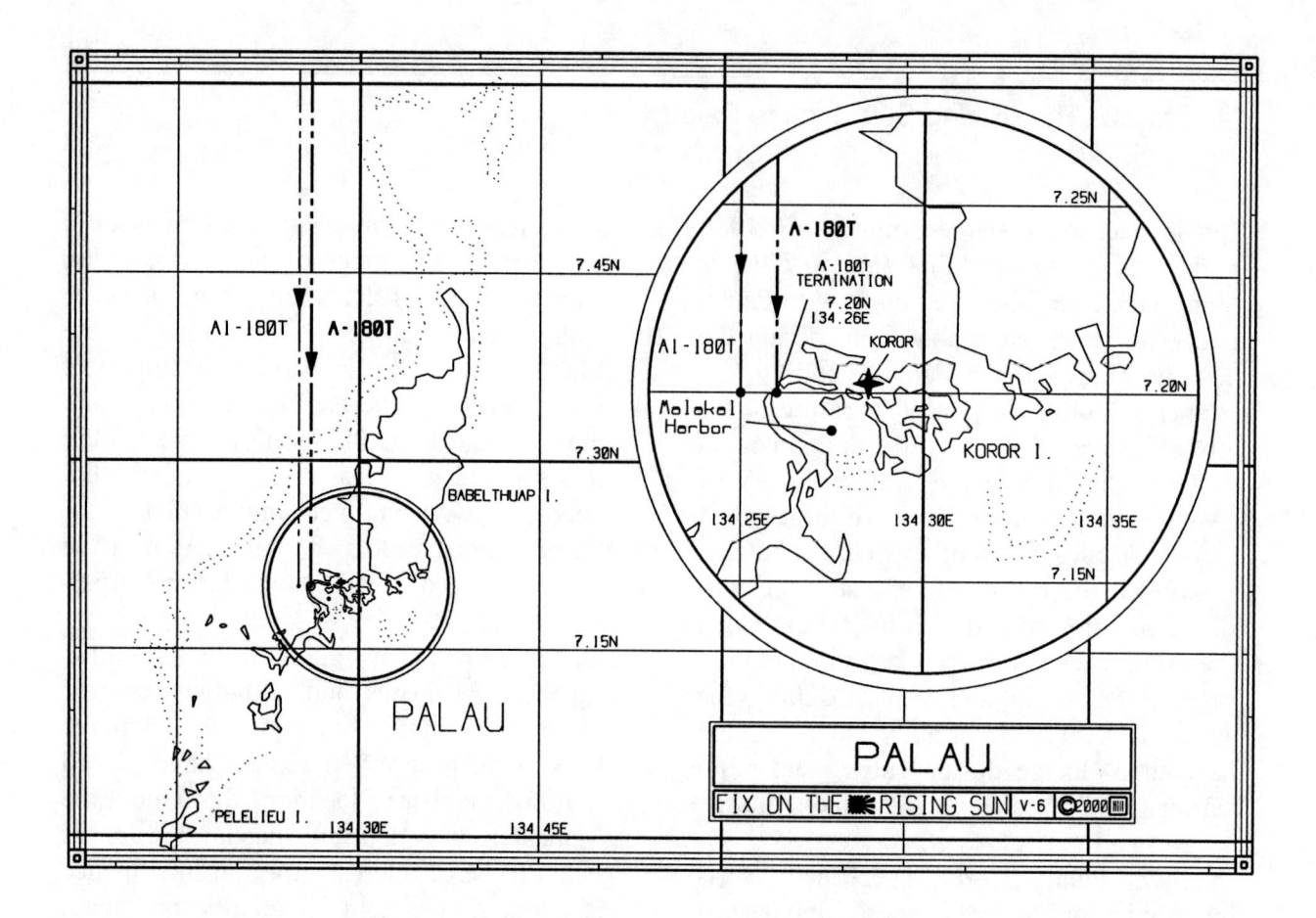

Hospitalized in Japan, apparently in a belief that immediate treatment might enhance his chances for recovery, Zembsch died when the Tokyo hospital was destroyed during the massive earthquake of September 1, 1923. Whatever he may have suffered in Koror, possibly due to his insistence on following orders to investigate Ellis' death, one can only imagine. But at least one report of his behavior, while in the hospital, indicates that Zembsch's passive detachment changed to restiveness whenever Japanese were present.

Moreover, despite the medical care provided to him, Zembsch was observed to cower in terror, whenever Japanese hospital personnel approached him directly, for treatment.

Clearly, the Japanese at Koror had intended that, regardless of any activities in which he may have become involved while at Palau, Zembsch was to serve as fair warning to the U.S., regarding espionage in the Mandates. Yet, the incident also indicates that the Palau Islands were not entirely isolated.

V-6, above, because of its scale, illustrates how the slight difference, between Pos. A and Pos. A1, greatly affects the significance of the Line of Position from the Clipper's 0200Z location. Pos. A, as reported, has a far greater significance than Pos. A1 (which had seemed more appropriate, at first, from a navigation standpoint). Of course, one questionable "LOP," in itself, is hardly proof of Japanese involvement in the Clipper hi-jacking. However, as will be shown, this is but one of *three* Lines of Position, which, together, establish the *Fix on the Rising Sun*, which indicates that the Clipper was, indeed, hi-jacked by elements of the Imperial Navy.

15. A SECOND "RISING SUN"-LINE: The Clipper's Fix on Saipan

The existence of graphic Lines of Position to Cavite and, apparently, to Palau, indicated that a search for another *Rising Sun*-Line might prove to be fruitful, and, as Pos. B and Pos. C had already been placed in doubt for dead reckoned navigation, they appeared to be the logical points at which to begin.

Pos. B appeared to have no immediately discernible function, but a Line of Position was calculated (not merely charted) from C to A and extended mathematically toward Saipan. The path of the extension, into the entrance to Saipan's Tanapag Harbor, was just too precise to be coincidental, especially since the LOP passed through the seaplane landing area at Saipan, as was also the case with the D-E LOP at Cavite and the A-180T LOP at Palau. The function of Pos. C, then, never valid as an example of navigation, was to provide a second "Rising Sun" Line.

While Palau served as the civil center for Micronesia and the Mandates, Saipan served as headquarters for the Imperial Japanese Navy. And, from the post-war account of a Chamorro woman, who had been eleven years old in the summer of 1937, it is known that the Chico-Tanapag Naval Base was then under construction. As the base included a seaplane facility, it can be assumed that it was active, at the latest, during the following year, in 1938, if not in late 1937.

Inasmuch as this young woman distinctly remembered both the year (1937) and the location, owing to the surprising appearance there of an *American flying lady* and her male companion, some may assume that she was "seeing things." However, a few PAA retirees have admitted taking an official, off-course, "peek" at this base in the late 1930's, so, it is likely that the base was over-flown during the search for *Hawaii Clipper*.

Therefore, because of a proximity to Guam and occasional inter-island civilian traffic, Saipan must also be ruled out, for obvious reasons, as a viable destination for the hi-jacked Clipper. However, the Saipan LOP, undeniably a *Rising Sun*-Line, does serve to define, along with the Palau LOP, a true (and significant) running fix for Pos. A:

Where a position fix is determined by at least two intersecting LOP's, established at separate times, the fix is said to be *a running fix*, that is, a fix established while running along a course. With two LOP's established, and centered on Pos. A, it may be said that they have established a *running fix*, of sorts, for Pos. A, and, since each was terminated at an important Japanese harbor, or naval seaplane landing area, it can be said, almost literally, that these two LOP's provided a running *"Fix on the Rising Sun."*

Initially, it seemed likely that the Clipper's navigator, Second Officer George M. Davis, had reconfigured Pos. A1 as Pos. A, for use as a due-south pointer to Malakal Harbor, and then established Pos. C as a pointer to Saipan. However, since he was under orders from the hi-jackers to provide a false, but reasonable, course to Pos. D, he moved Pos. C, his Saipan pointer position, precisely 1 degree to the west, placing it at Pos. C1, which had the superficial appearance of a valid position between Pos. B and Pos. D. It remained only for FRO William McCarty to substitute just one digit in the Longitude for Pos. C1, while actually transmitting Pos. C. Pos. C1 certainly appeared to have been sent, which satisfied the hi-jackers, but it was Pos. C which was actually received at Panay by radio operator Fernandez, thus preserving the pointer to Saipan, within a clearly spurious position report, for anyone at PAA who re-plotted the route data.

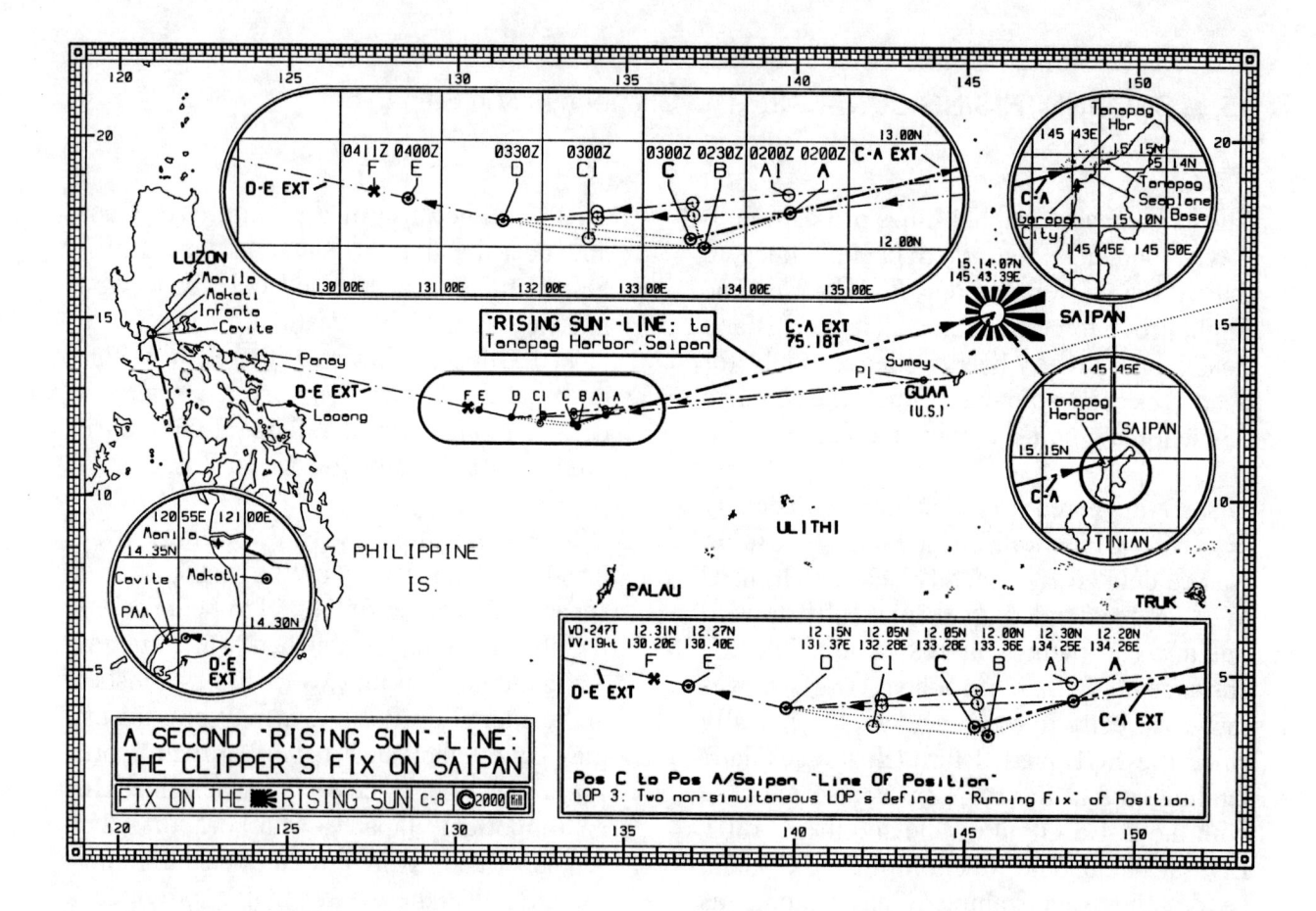

That the hi-jackers depended upon the PAA Clipper crew to provide both navigation and communication services, was, by no means, unlikely: the two hi-jackers would certainly have had their hands full just maintaining control of the *Hawaii Clipper*'s passengers and crew. Because maritime navigation is a graphic art which can be evaluated, if only in a superficial manner, simply by looking over the navigator's shoulder, it is not at all improbable that Second Officer Davis was put to work. Moreover, by 0200Z, Davis had had six hours and twelve position reports, since departing from Guam, to enhance his credibility as a "cooperative" hostage, and the hi-jackers had endured increasing stress during the same period. The result may have been that the inspection of Davis' work was gradually diminished as the day wore on.

As to the hi-jackers allowing FRO McCarty to continue providing communications, this would have been a requirement, rather than a convenience: professional radio operators invariably have a distinctive "swing" to their telegraphy—an individual style which is easily recognized by other operators. Had McCarty been replaced by another operator, then Edouard Fernandez, at Panay, would, without any doubt, have picked up on the substitution, and the jig would have been up.

Although the Saipan LOP (with the Palau LOP) defines the running fix, it is still one of only two LOP's which seem to implicate Japan in the hi-jacking of *Hawaii Clipper*. However, navigators have a propensity for ensuring the validity of a fix by establishing at least *three* valid LOP's, not merely two, in which case the fixed position may then be considered to be entirely reliable. Of the two doubtful positions, Pos. B and Pos. C, only Pos. B remained to be tied in, and, in fact, it seemed chronologically reasonable that the Saipan LOP might, in fact, have been the third, not the second, *Rising Sun*-Line.

16. SAIPAN: The "Rising Sun"-Line to Saipan

On June 15, 1944 (June 14 in the U.S.), U.S. Army and Marine forces landed on Saipan's west coast, their invasion waves centering on Charan Kanoa, south of Garapan City, and encompassing almost the entire southern coast. This was the "D-Day" of the Pacific War, presaging an ultimate U.S. victory, but it was eclipsed by the Normandy invasion which had occurred only eight days before, even as the Saipan anniversary would be eclipsed by the Normandy commemorations, fifty years later.

However, the great import of the invasion of Imperial Japanese Navy Headquarters was not lost on Japan, and, according to some accounts of the evening of the first day, Tokyo Rose is supposed to have played a recording of John Phillip Sousa's *Stars and Stripes Forever* during her regular evening radio broadcast on *The Zero Hour*—and to have saluted the American accomplishment.

But what began with one long day was not soon finished. It took two weeks of savage fighting to reach Garapan City, and it hadn't yet ended by July 9, when American troops approached Marpi Point and watched, with horror, as hundreds of Japanese soldiers and civilians, convinced by Japanese propaganda that the American G.I.'s would kill them if they surrendered, leaped to their deaths from the north coast cliffs.

Saipan was full of surprises. Four days after the invasion began, two soldiers attached to a U.S. Army Postal Unit watched as Amelia Earhart's Lockheed *Electra* was flown once from newly captured Aslito Airfield—then later burned on the orders of Secretary of War Forrestal—a claim which, forty years later, prompted one of them, Thomas Devine (who then lived not far from the home of Amelia Earhart's sister, Muriel), to write a book about his claim. Curiously, one of the larger sections of the U.S. Navy's Earhart File outlines the Navy investigation of Tom Devine—and its attempts to discredit him.

Two weeks into the Saipan invasion, U.S. forces captured Garapan City, where an *American flying lady* had been put up in a room at a local hotel in 1937—and where her injured male companion had been seen in custody at the city jail. A few days later, the Americans captured Tanapag, where, in 1937, an eleven-year-old girl had seen an American man and woman emerge from a beached aircraft at the new seaplane base. Many Americans, then as now, believed that both Earhart and Noonan had fallen into Japanese hands, and so, eyewitness accounts were not unexpected after the war. But, in 1944, another surprise, entirely unexpected, came to light at the Garapan City bank:

The following is the text of a note written by Joe Gervais, the Earhart hunter, after he had heard the account of this Saipan "surprise:"

"During the invasion of Saipan during WWII (the Southwest Headquarters for the Imperial Navy) CWO W. T. Horne, U.S. Marine Corps, now retired, stated that, in the city of Garapan on the island of Saipan, stood the vault of the bank, surrounded by reinforced concrete, despite the entire bank's having been levelled by shelling and bombing. Horne, with a "swat team" of Marines carrying satchel-charges, set the charges and blew the main door off the vault. Inside they found ten million dollars in neatly packed bundles of specially printed invasion money to be used during the occupation of the Hawaiian Islands. Also found was three million dollars in U.S. gold-backed paper currency of the pre-WWII era. The entire contents of the vault were turned

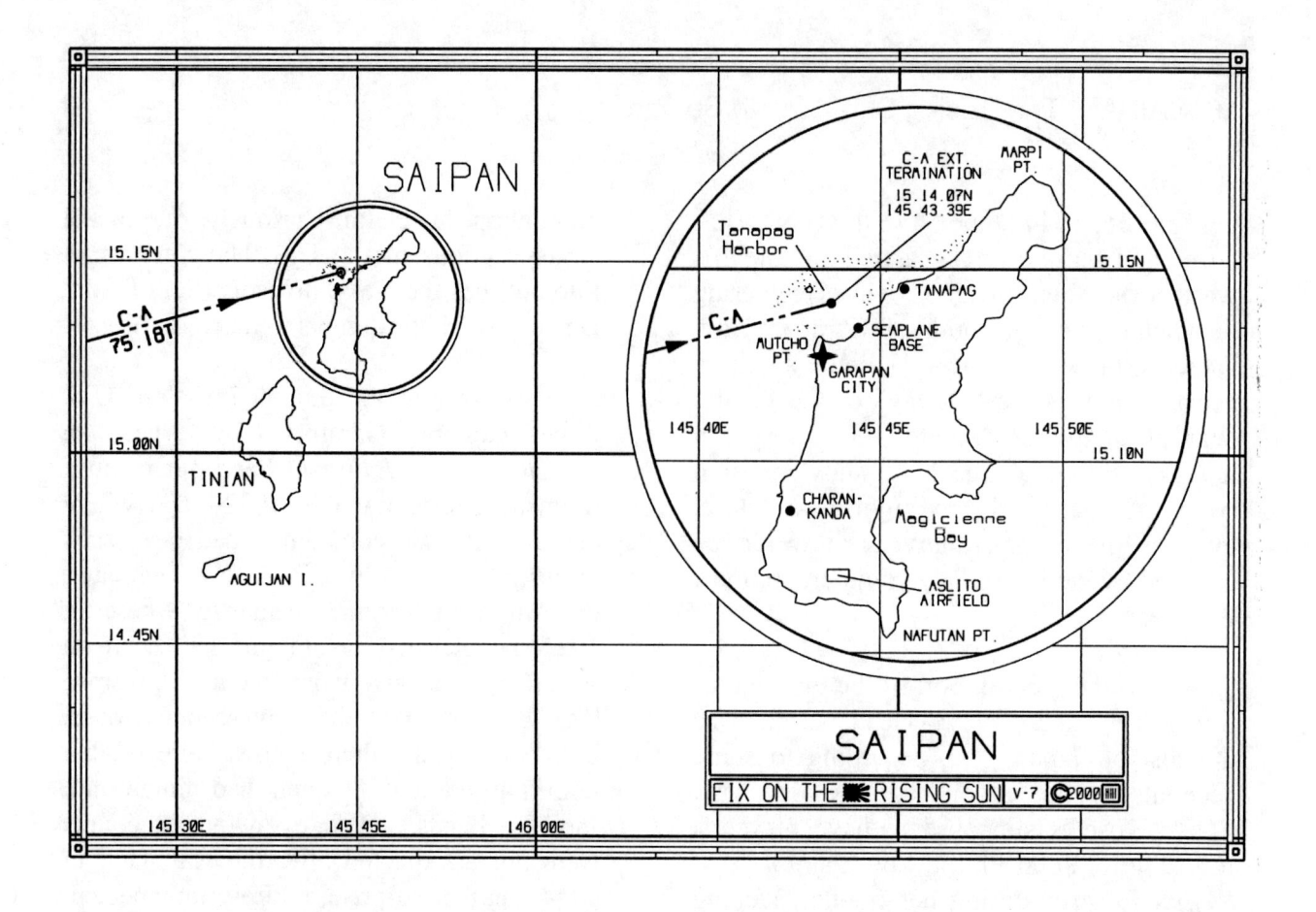

[Map of Saipan with inset detail, labeled "SAIPAN — FIX ON THE RISING SUN v-7 ©2000"]

over to a Marine Intelligence Unit headed by a Colonel. (Is this the money obtained from the Chinese businessman during the hi-jacking of the Hawaii Clipper in 1938?)"

If this were, indeed, the "war relief" money which Wah Sun Choy was carrying to China aboard *Hawaii Clipper*, the U.S. Treasury could have identified it, as Gold Certificates were certainly not "pocket money" after 1933. This would not have been the first inkling in Washington, that the Japanese had hi-jacked *Hawaii Clipper*, nor would it be the last. And yet, the American government suppressed all evidence of the hi-jacking, and continues to do so in deference to Japan.

But, for those who cannot bring themselves to believe that Japan wields sufficient power to compel America to cover up Japanese "incidents" of more than half a century ago, perhaps a footnote to the Saipan invasion may prove enlightening:

For more than two weeks prior to the fiftieth anniversary of *D-Day*, the heroic Normandy Invasion, America was subjected to a heavy media "blitz," culminating in the coverage of President Clinton's visit to the Normandy beaches on June 6, 1994. However, veterans of the Saipan invasion—which was no less costly or *meaningful* than the Normandy invasion—may recall that there was almost no mention of the Saipan invasion of June 14, 1944 (the U.S. date), in America's press.

They may *not* recall that on June 13, 1994, the eve of the Saipan invasion anniversary, the President of the United States, William Clinton, entertained, at the White House, not American veterans of the Saipan invasion, but *Akihito, Emperor of Japan*, who spent that night at the White House and who only departed for Japan *after* the very day when, fifty years earlier, Japan had conceded the American victory at Saipan. Today, the U.S. concedes the Japanese victory over *history*.

17. A LAST "RISING SUN"-LINE: A Speed-Made-Good Fix on Truk

Because the Palau LOP, A-180T, appeared to be the first of the *Rising Sun*-Lines (at 0200Z) and since the Saipan LOP, C-A, was surely the third (at 0300Z), then it seemed likely that an LOP involving Pos. B might provide a *second* LOP (at 0230Z). It should be noted that a simple change of only a single degree or minute of Longitude (C1-C, E1-E), had permitted the transmission of a reconfigured false position by substitution of a *single digit* during transmission of a report.

By moving Pos. B only *two* degrees to the east, Pos. B1 was created, which still met the requirement for a simple, single-digit substitution during transmission of a report. As with the Saipan LOP, a course was laid from A to B1 and mathematically extended. Despite the distance, this course not only crossed Truk Atoll, but passed within about 2.5nm of the center of the Imperial Navy's Fourth Fleet anchorage: Eten Anchorage.

Because Truk, for all its isolation, seemed a logical destination for the hi-jacked Clipper (especially in light of Gervais' story about the entombment of the fifteen Americans on Truk, in 1938), it appeared quite reasonable to examine this course in some detail. As the inset illustrates, the course from A to B1 very nearly meets the requirement for a TAS of 128kt, but it does not meet the necessary requirement that a part of it lie on or near the "course-defining" Makati bearing of 101T.

The extended Truk LOP, A-B1, could not, therefore, have been the actual final course of *Hawaii Clipper*, but it appears to provide the third (actually the second) valid *Rising Sun*-Line, and, with the Palau LOP, A-180T, and the Saipan LOP, C-A, *confirms* the *running fix* for Pos. A and also confirms the implication of the Imperial Japanese Navy in the hi-jacking of *Hawaii Clipper*.

At this point, it may be useful to review the few positions already discussed, especially before proceeding to a presentation of the *actual* hi-jack route from Guam to Ulithi and Truk. By conventions adopted for this work, *true positions* are those actually attained by the Clipper. *False positions* are those along the false route and unattained by the Clipper, but which are flight-valid, for dead-reckoned navigation. The *reconfigured (false) positions* are those modified by the Clipper's navigator for the specific purpose of generating the *running fix* LOP's:

Pos. A was quite likely a true position, accurately reported, despite the hi-jackers' apparent original plan to reach Pos. A1. As will be seen, Pos. A, certainly the work of a "helpful" George Davis, restored the much-delayed hi-jacking schedule, but also provided a basis for Davis' LOP's.

Pos. A1 was, apparently, a true position in the hi-jackers' *original* operational plan, but it was replaced by Pos. A, (with Davis' help) due to unexpected flight delays.

Pos. B was a false position, contrived by George Davis to provide a reasonable 0230Z position (to satisfy the hi-jackers), but also selected to permit Pos. B1 to be transmitted by digit substitution, to provide a pointer to Truk. However, Pos. B, instead of Pos. B1, was actually received by Panay, for the reasons noted in the following paragraph.

Pos. B1, was a reconfigured false position derived from Pos. B, but it may not have been transmitted at all, due to its unusual placement, which may have alerted the hi-jackers. Or, it was sent, but questioned by Fernandez at Panay, who may have asked for, and received, a correction, Pos. B, there being such a risk in re-sending Pos. B1.

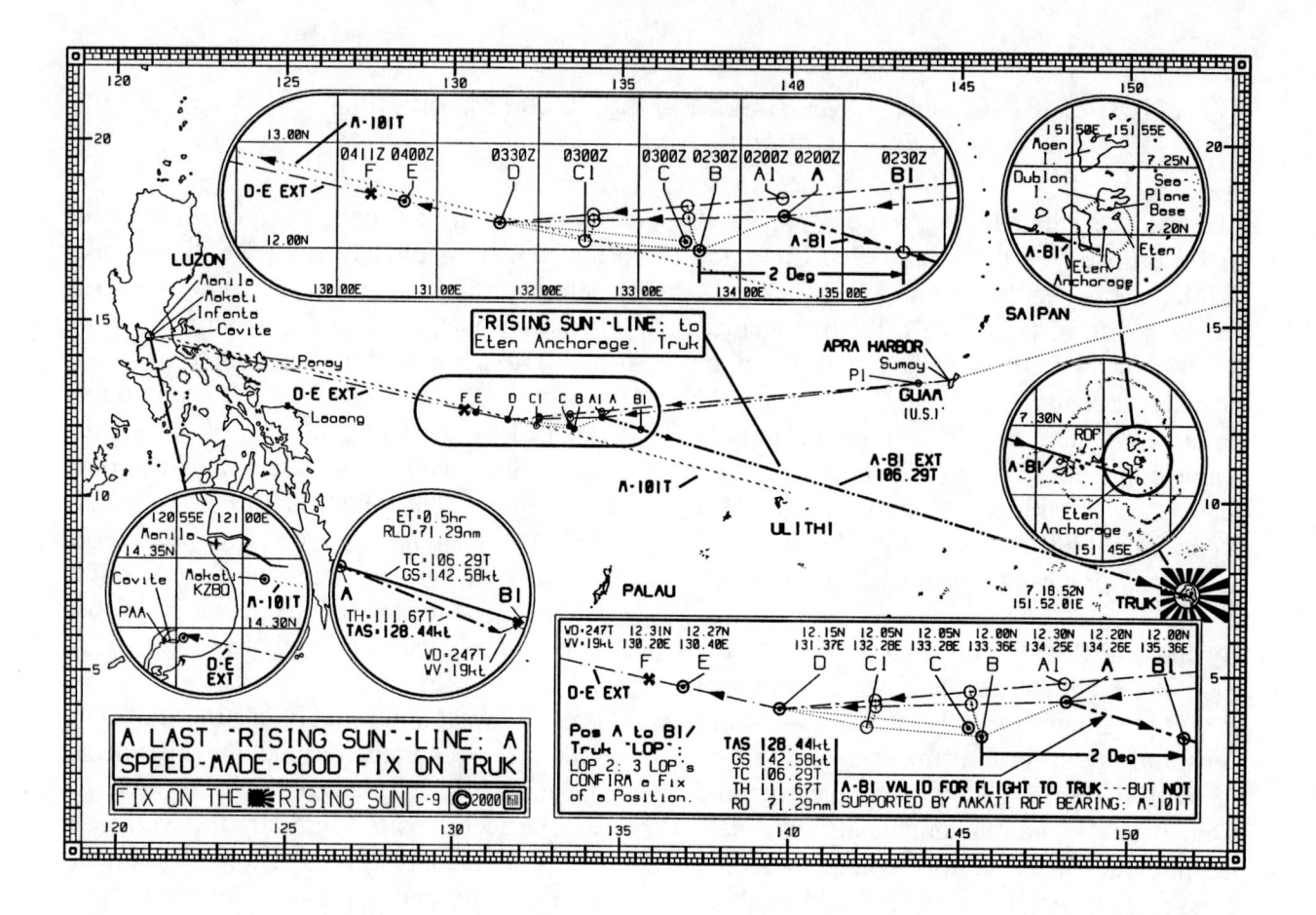

Pos. C was a reconfigured false position, derived from Pos. C1, and was sent by digit substitution, thus providing the Saipan LOP, or pointer. Despite its unsuitability when plotted, with Pos. B and Pos. D, it was not so obviously out-of-place as Pos. B1 and so, was probably not challenged by Panay.

Pos. C1 was a false position, contrived by the navigator to represent a reasonable 0300Z position (to satisfy the hi-jackers). It was reconfigured as Pos. C, which, as noted above, was successfully sent by digit substitution, defining the Saipan LOP.

Pos. D was a false position, planned and selected, by hi-jack planners, to serve as the timing and positional linch-pin, both for the hi-jacking and for all the peripheral support activities. Pos. D was the only position which could not be reconfigured by Davis, but he used it to create his own D-E LOP, as well as to serve the hi-jackers' needs.

Pos. E was a reconfigured false position, derived from Pos. E1 and transmitted by digit substitution to express Course Made Good from Pos. D and, thus, to provide the graphic D-E LOP extension to Cavite. This served as the "key" or "legend" to aid the decipherment of Davis' "message."

Pos. E1 was a false position selected by the hi-jack planners to provide a reasonable position by which to confirm Speed Made Good from Pos. D (with Course Made Good to be confirmed by the 101T bearing from the Makati RDF). Pos. E1 was never sent, but was reconfigured by Davis and sent out as Pos. E, using digit substitution.

Pos. F was dead-reckoned for this work from the last data received from the Clipper. It indicates the Clipper's false "position" at 0411Z, when radio communication was lost. Pos. F should not be confused with the oil slick origin, further along the false course.

18. TRUK: The "Rising Sun"-Line to Truk

Although Truk had quite a large indigenous civilian population scattered over numerous islands within the surrounding reef—with a history of inter-island trade dating to the nineteenth century—the development of the Japanese naval base, at Dublon Island, after 1915, had necessitated a far greater degree of security for the atoll than was necessary, or even possible, at either Palau or Saipan.

Furthermore, Truk was so remote from any American territory and so well protected by its surrounding reef, that it was unlikely that any American ship, submarine or aircraft could penetrate the area to search for a lost airplane. Consequently, Truk was the ideal place for the *Hawaii Clipper* to be initially delivered until an ultimate destination could be determined. In addition, Truk had been an ideal location for the Fleet Faction cadre to congregate and to operate: Truk's remote location kept it especially independent of Japanese who were distrustful of the militant Fleet Faction—as echoed today, perhaps, in the tension which separates the U.S. Navy from its so-called "Submarine Mafia."

In this regard, it should be remembered that, while Japan surrendered, both as a nation and as a military power, in September, 1945, the garrison at Truk refused to surrender until February of 1946, nearly six months later. And this wasn't simply a case of a few fanatical holdouts, hiding in the bush, but of a nearly official mutiny by a sizeable, well equipped naval and military force, easily capable of exacting a terrible price for its defeat. In 1945, as in 1938, Truk was both so isolated and so powerful as to have been quite an independent power, in itself.

Historically, the Japanese presence on Truk pre-dated by decades the League of Nations Mandate and was firmly founded upon the establishment, in the late 1800's, of several successful Japanese trading companies. But after the naval base was created, in 1915, relations with the indigenous Micronesians were not inclined to endear the Japanese overlords to the civilian population. Basic education was not permitted to progress beyond the fifth grade level, and disrespect to a Japanese could produce any punishment, from a severe beating to death by beheading. On the positive side, the Japanese saw to it that there were jobs for the civilians, and so, there came to be some measure of prosperity under the League of Nations Mandate.

Obversely, with the arrival of the Americans after the Pacific War, education was made possible to about any level attainable by the student, but the Micronesians were left to fend for themselves, economically, and so, a grinding poverty settled over Truk—even as it did elsewhere within the former Mandates. Since 1986, Truk Atoll (now known as Chuuk State) has been a State within the sovereign Federated States of Micronesia, and it has begun to enjoy a resurgence of Japanese interest, by way of trade and tourism. The continued American presence (with financial aid) assures both education and protection for individual liberty, and so, Truk now benefits from both influences. It is unlikely, therefore, that Truk State will welcome any attempt to dig up portions of the past which might embarrass either—or both—of its present benefactors.

The Clipper is no longer at Truk, of course, having been flown to Japan, probably within a few days following the hi-jacking. After WWII, General MacArthur's chief of staff, General Richard Sutherland, informed Edna Wyman, widow of passenger Ted Wyman, that the Clipper engines had been recovered in Japan, possibly at Gifu, near Nagoya.

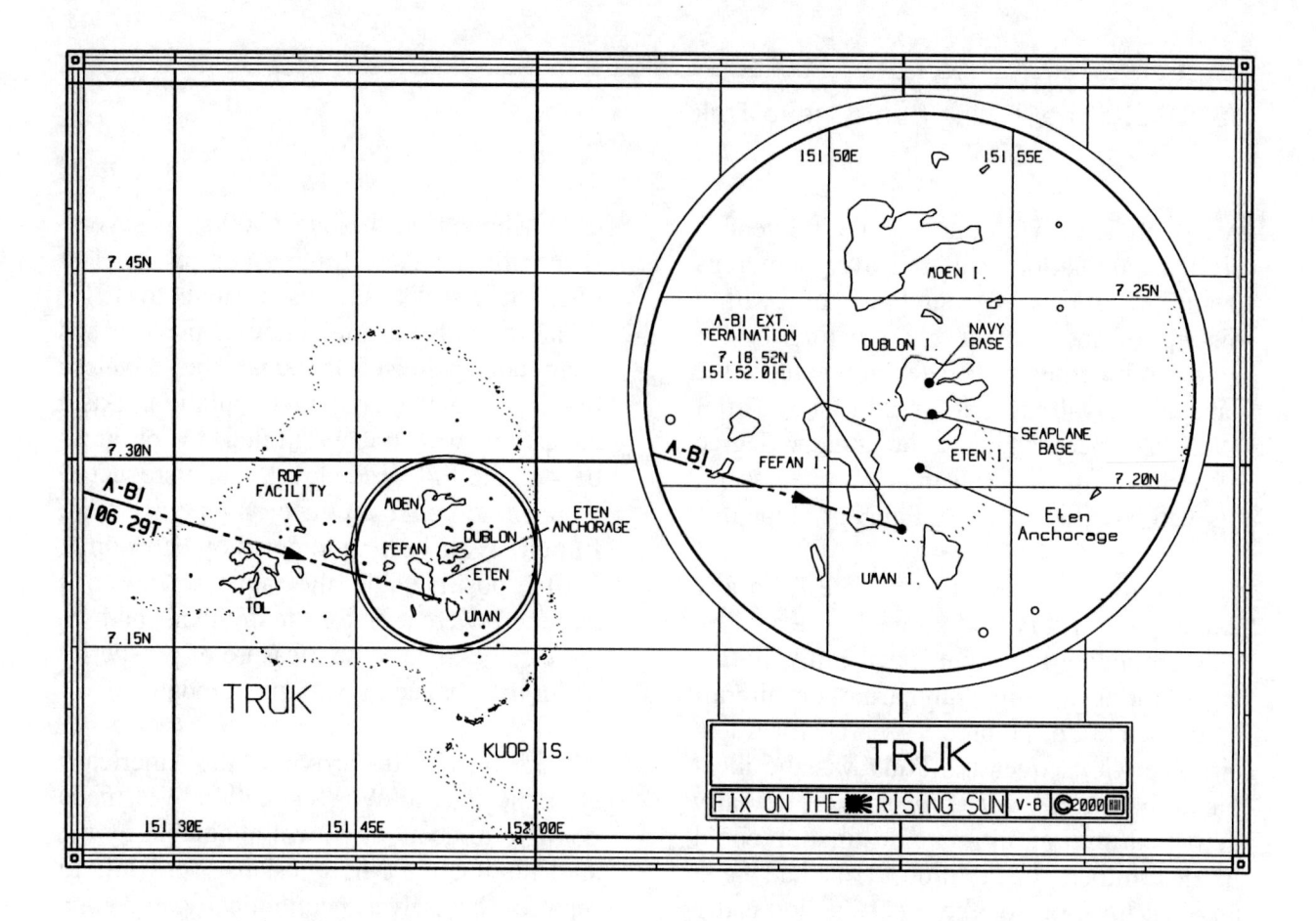

But there are substantial indications that Ted Wyman and his fourteen fellow air travelers still remain on Truk Atoll.

The report on the fate of the fifteen crew and passengers will be presented in a subsequent section, but that report has suffered from its association with a story which indicated the presence of *Hawaii Clipper*, itself, on Truk:

In 1964, during an investigation of an aircraft site on Dublon Island, Joe Gervais, the Las Vegas Earhart hunter, was informed by his local guide, a former contractor to the Japanese by the name of Robert Nauroon, of Nauroon's involvement in the entombment of the Clipper victims. In following up on Nauroon's tale, Joe was shown the wreckage of a flying boat, which he initially thought might be the lost Clipper. Later, upon returning to the United States, Joe released, to the Associated Press, both the Nauroon report and his wreckage snapshots.

Both aspects of the story, including the tale told by Nauroon and Gervais' photographs, were reported by the *New York Times* and the *Los Angeles Times* on July 6, 1980. But, because the wreck was subsequently shown *not* to be that of a Martin *M-130* flying boat (that is, not the *Hawaii Clipper*), the report from Robert Nauroon was also cast in doubt.

On July 14, Ted Wyman's daughter wrote to Pan Am, requesting that Joe Gervais's story be verified, and on August 8, James Arey, Director of Corporate Public Relations for Pan Am, responded by casually rejecting Nauroon's report, primarily on grounds of the erroneous identification of the wreckage.

But the greatest barrier to truth still remains at Truk—the Amelia Earhart Law, which forbids private searches for human remains. Oddly enough, as the author is informed, the name "Amelia Earhart" was quite common among the Trukese, prior to the Pacific War.

19. THE RADIO BEARING *MIRROR*—A Brilliant Radio-Deception Plot

During the first half of the 20th Century, there were three radio-deception plots of historical moment, all involving naval radio bearings. The first of these was perpetrated by Germany against Russia in the Black Sea, early in 1915, during the Great War—WWI. The third was perpetrated by Russia against Japan, on behalf of the United States, in the western Pacific, in late November and early December of 1941, just prior to the attack on Pearl Harbor. The second was perpetrated by Japan against the United States in 1938, in the Philippine Sea just west of Guam— the key to the hi-jacking of *Hawaii Clipper*:

The Japanese used a radio-deception method that can only be described as a radio bearing *mirror*. It functioned by playing on PAA's unwarranted faith in the *single* bearing as a means of confirming normal Clipper flight operations—and yet, the single bearing can only establish a presence *along* the bearing. It cannot, by itself, establish a *position* fix, that is, a specific position along the bearing. *Two* bearings are required, with the position fix lying at the point where the two cross.

C-10 depicts the apparent hi-jack plan. Additions to the plan appear on C-11, and what actually occurred, when Japanese plans went very wrong, is shown on C-12.

The plan turned on a planned 282000Z take-off, after which, at 282145Z, the hi-jackers would have made their presence known and taken command. A turn south at 282200Z would have brought the Clipper to Pos. A1 at *290130Z*, at which point the Clipper would have turned east toward Pos. D1. Intermediate (true) positions en route to Pos. D1 would have been paired with apparent (false) positions, which, when transmitted to Panay, would give the impression that the Clipper was still en route to Pos. D, while actually en route to Pos. D1. The Clipper would fly from D1 to E2, en route to Ulithi, while Pos. E1 would establish an apparent (but false) course from Pos. D to Cavite. All the positions were selected as *mirror-pairs*, which had to satisfy four parameters:

First, the intended positions had to produce a dependable route to Ulithi, while apparent positions had to indicate a viable, but false, route to Cavite. *Second*, both routes had to reflect a TAS of 128kt, to conserve fuel and to provide for standard *M-130* operations. *Third*, both the intended and apparent routes had to satisfy Wind Triangle calculations, the former to ensure arrival at Ulithi and the latter to establish a viable false route. (Note: because the actual winds were westerly, the intended route would cover more distance than the apparent route, since it enjoyed an effective tail-wind). *Fourth*, both positions of each mirror-pair had to lie on a common radio bearing from Makati. That is, Pos. D and Pos. D1 had to share virtually the same radio bearing from Makati (101.37T), as did Pos. E1 and Pos. E2 (101.51T). So long as the flight appeared normal, no RDF bearings would be attempted from Sumay to establish a fix on the Clipper's true position. Makati would assume that the Clipper was, indeed, at Pos. D, when in fact, it was transmitting from Pos. D1, and this would hold true for all mirror-pairs from A1 to D and A1 to D1.

Because the winds could not be predicted in advance of the operation, the two hi-jackers undoubtedly had a number of optional flight plans, each of which would have been based on a key control point, such as Pos. D and Pos. D1 (in this particular case). In fact, so long as the control points and their arrival times were specified, all the rest of the intended and apparent positions could simply be left to the Clipper's navigator.

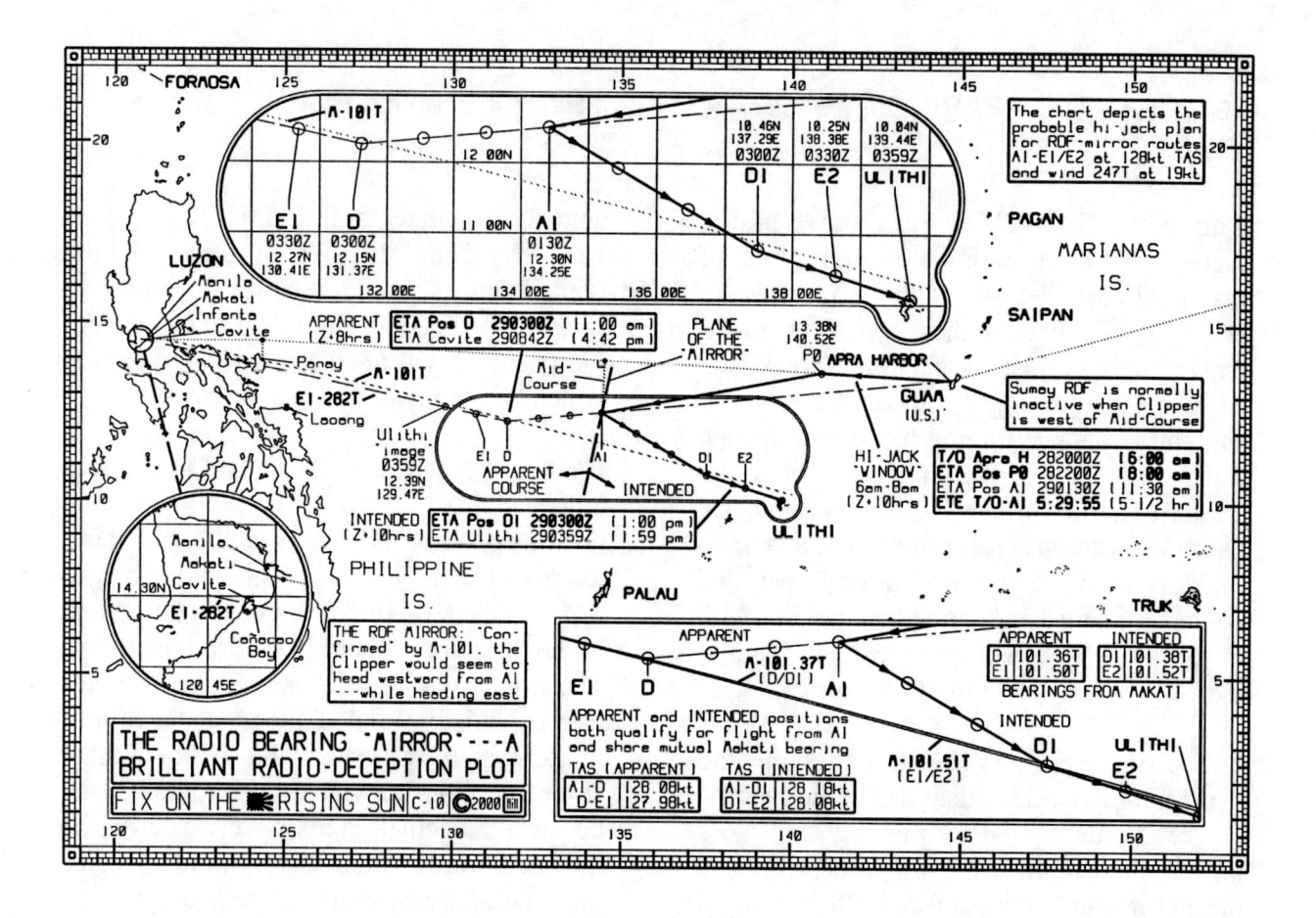

It should be noted that, while the hi-jackers derived Pos. D (apparent) from Pos. D1 (true), Pos. D1 had to be derived from Pos. D for the purposes of this analysis. In fact, there is, within normal limits, no position, other than Pos. D1, which will satisfy the specified parameters in relation to Pos. D, and vice-versa. The same applies to Pos. E2 and Pos. E1, and, as will be seen on C-11, these critical control point arrival times had to occur *on the hour or half-hour*, to ensure that these specific points would be reported.

Ulithi was quite a large atoll, but one which enclosed no large islands. The indigenous people of Ulithi were (and still are) fiercely independent, and, perhaps for both of these reasons, the Japanese never fully developed Ulithi, as they did other island groups in the Pacific. However, by 1938, they had built a small radio station and a seaplane service facility at the atoll, and the hi-jackers had certainly planned to drop a relief flight crew there, prior to the hi-jacking. It is likely that the crew of *Hawaii Clipper* were to be told, or, led to believe, that they were being hi-jacked directly to Truk and that Ulithi would provide only a navigation checkpoint for that route. Such a belief could be expected to cause the PAA crew to act cautiously, at least until they had reached what they would believe was only a checkpoint, where stress fatigue could be expected to take its toll on the hi-jackers, and resistance might begin. By planning to drop into Ulithi with little warning, the hi-jackers could minimize the risk of early resistance—until it was too late for resistance—and the flight to Truk could be completed with a fresh Japanese crew.

It seems likely that the hi-jackers planned to use explosive devices as a means of forcing the Clipper's crew and passengers to accept the hi-jacking without opposition, but such a threat would have been effective only after take-off, when the Clipper was in the air.

20. ADDED DIMENSION: The Shadow of a New Player Crosses the Board

Admittedly, C-11 is essentially hypothetical and is supported by circumstantial evidence. However, C-11 offers answers to a number of valid questions, which are germane to the last flight of *Hawaii Clipper*. C-11 indicates the presence of another player in the game, a Japanese Navy flying boat, and postulates that *Hawaii Clipper* had been targeted for an air attack, had the on-board hi-jackers failed.

Thursday afternoon, July 28, as the Clipper landed at Apra Harbor from Wake Island, an armed four-engine Kawanishi *H6K* flying boat would approach Ulithi from Truk. An improved copy of Sikorsky's *S-42*, the *H6K* was not entirely well suited to long Pacific flights, but Japan was able to make good use of these aircraft throughout the Mandates.

This *H6K* would be carrying an additional flight crew, who would disembark at Ulithi, even as the two Japanese hi-jackers prepared to stow away, that night, aboard the Clipper, at Apra Harbor. At 2000Z the next morning, as *Hawaii Clipper* took off for Cavite, the *H6K* would take off for Malakal Harbor at Palau. Her radio operator would monitor all of the Clipper's reports and, by 2314Z, as the *H6K* landed in Palau, her Captain would know which mission he was to pursue:

I. Hi-jack Support Mission: If intercepted Clipper flight reports revealed that *Hawaii Clipper* was en route to Pos. A1, the *H6K* Captain would know that the Clipper had been successfully hi-jacked. He would then order that two specially constructed oil drums be loaded on board the *H6K*. These were to contain a mixture of very light oil and gasoline and be weighted to sink upright to the bottom of the sea. Vents on the drums would open under pressure, soon after they were dropped, ensuring that the gasoline/oil mixture would seep out for several days.

After refueling during a one-hour layover, the *H6K* was to take off at 0016Z and proceed at 128kt TAS to Pos. D, arriving at 0300Z, when she would turn on to the D-E1 course to Cavite, as the hi-jacked Clipper turned onto the D1-E2 course to Ulithi. Fifteen minutes from Ulithi, at *0344Z*, the Clipper's radios would be silenced. *Hawaii Clipper*'s crew would suddenly be dismayed to learn that they were landing at Ulithi, where at 0359Z, the waiting Japanese flight crew would take command and relieve the two hi-jackers. Also at *0344Z*, the *H6K*, on the false mirror-route, would drop the oil drums, creating a deceptive oil slick.

Then, if the weather in the Philippines was overcast, the *H6K* would proceed along the D-E1 EXT at TAS 128kt and, after making a brief "appearance" at about 0700Z, high over Luzon, would turn toward the Japanese seaplane base at Takao, Formosa. If the *H6K* were heard above the overcast, or were even glimpsed through breaks in the clouds, these two deceptions might serve to divide the search for the Clipper. Essentially, it was a simple plan, involving only limited cost and providing for the failure of the hi-jacking:

II. Intercept Mission: If, during his morning flight to Palau, the *H6K* Captain determined that *Hawaii Clipper* had *not* turned onto the course toward Pos. A1, but had continued, instead, to proceed on the Direct Route to Cavite, he could assume that the hi-jackers had failed to board or to hi-jack the Clipper, and he would then prepare for his alternate mission, after landing at Palau at 2314Z:

In such a case, there would be little need for the oil drums—and little time to load them. The Captain would have to refuel quickly and take off at 0000Z on the route from Malakal Harbor, Palau to Takao, Formosa.

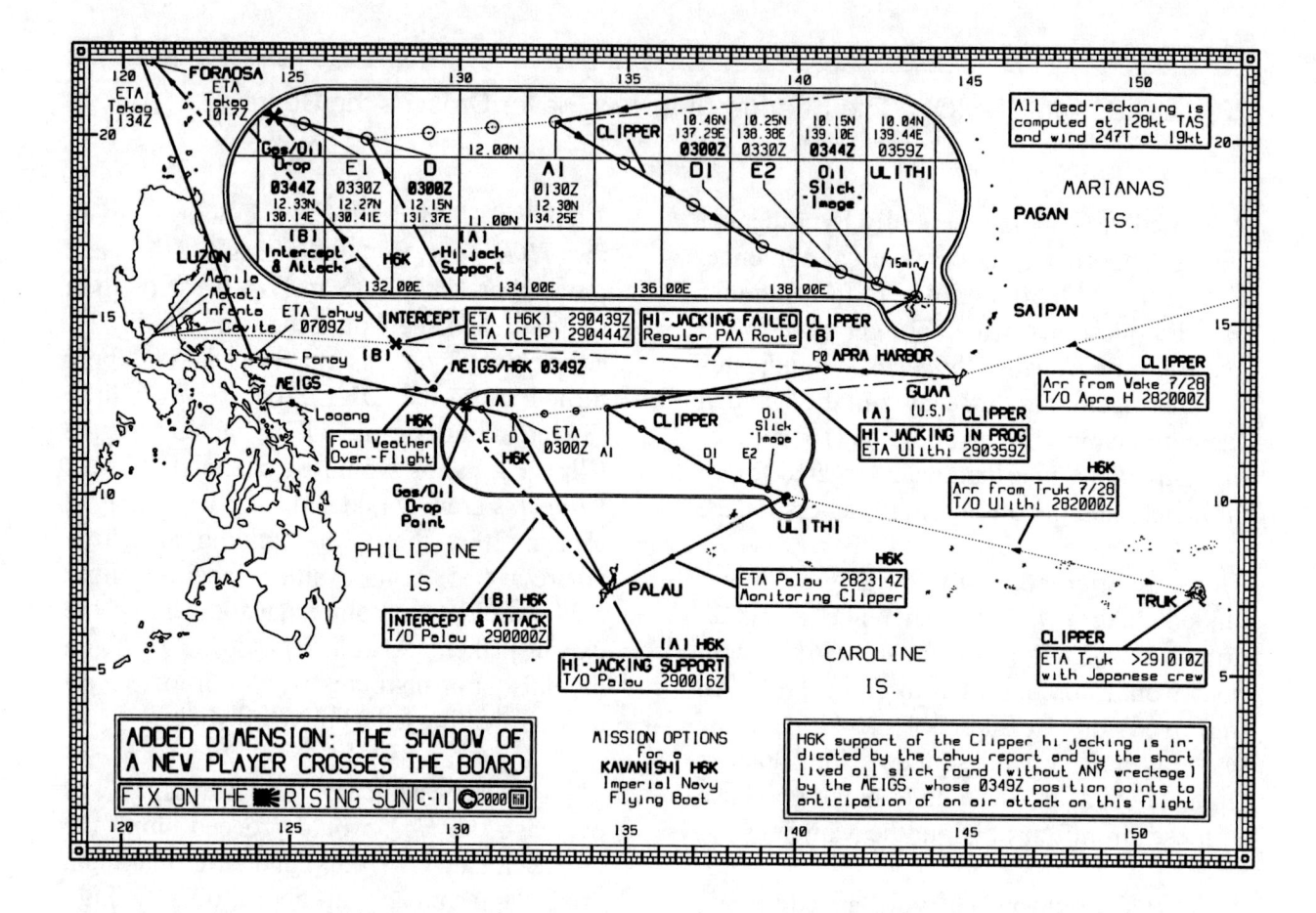

His objective would then be to intercept the Clipper, and he could reach the intercept point at 0439Z, five minutes ahead of the *Hawaii Clipper*. During the quarter-hour prior to interception, he would have used the Clipper's 0430Z flight report and his own loop RDF to locate the PAA Clipper and to determine a vector from which to launch an attack. At about 0444Z he would dive on the Clipper, either from out of the sun or from out of the clouds (depending upon weather), raking the starboard side of the Clipper's bridge with machine-gun fire to silence the radio—and the Flight Radio Operator. Then, if sea conditions permitted and the Clipper pilot responded properly to visual signals from the *H6K*, the Japanese flying boat would follow the Clipper down to the sea. If the Clipper landed successfully, the *H6K* Captain, after weighing all the risks, might land, as well, and attempt to board her and resurrect the hi-jacking, using the original hi-jackers, if alive, or else one of his crew.

Of course, at any time after the intercept, if any resistance were observed, or even heard (from the Clipper's emergency transmitter), the *H6K* Captain would simply finish off the Clipper and resume his flight to Takao. But Japan needed one the Clipper's new engines, so finishing her off would be a last resort.

The U.S. had anticipated just such an attack: Warnings had surfaced in June, and it could hardly be mere coincidence that, at 0349Z, USAT *Meigs* was only a mile or so west of the point where the *H6K* would cross the Meigs' course to Guam, during its intercept flight to Takao. Had the Clipper not been hi-jacked, *Meigs* was in an excellent position, at that time, to see (or to hear) the *H6K* overhead and to warn PAA of an impending intercept. This would explain why Panay was able to establish radio contact with the *Meigs* so quickly, at 0430Z, soon after the Clipper's radio became silent (at 0411Z), but long before she was presumed lost.

21. MALICE THROUGH THE LOOKING-GLASS: The Clipper Comes About—for Truk

C-12 defines the events that attended the final flight of *Hawaii Clipper*, as supported by the available data. The following data, as gleaned from PAA personnel, perhaps in bits and pieces, by determined AP reporters, no longer exists in official files—but cannot be, and *was not*, disregarded in this analysis:

July 30, 1938, the *New York Times* reported: "Half-hourly reports from the Clipper after lifting from the water at Guam at 4:08 P.M. New York time [2008Z] gave no indication of trouble, according to the Associated Press. The 5 o'clock report [2100Z] showed that the boat had climbed to 10,000 feet and attained a ground speed of about 105 knots an hour [sic]. Regular reports from that time until the last message at 11 o'clock [actually 12 o'clock, or, 0400Z] showed the ship to be cruising at between 105 and 112 knots at altitudes between 10,000 and 9100 feet."

Hawaii Clipper could hardly have flown directly to Pos. A. As shown on C-4, her ground speed would have been 102.49kt, yet the *NYT* piece had noted a minimum ground speed of 105kt. To dead-reckon a 5 hour, 52 minute route from Apra Harbor to Pos. A at 105kt GS, required a steady course on the Direct Route to Cavite until 0000Z, at Pos. P8, and thence to Pos. A on a course which, in fact, approximates the reported course line from A to B. This yields an ETA at Pos. A, as reported, at 0200Z. The dead reckoned time on this route, at 105kt GS, is in error by only *1 second* (as seen on V-10), indicating that this was the true route followed.

It must have been quite a surprise to the hi-jackers, emerging from hiding at 2145Z, to find, not only that the Clipper had taken off 8 minutes late, but that she had been flying at 105kt GS (122.4kt TAS), rather than the planned speed of 110.62kt GS (128kt TAS). The slow speed had been specifically chosen to upset the suspected Japanese plans for an air attack, but it also upset the hi-jack timetable, as well, and required that the hi-jacked Clipper kill time, a full 22 minutes, en route to Pos. A1 (soon to be replaced by Pos. A).

Things were not going well for the *H6K*, either: With the Clipper still maintaining her regular route to Cavite, prior to 0000Z, and the *H6K* ordered to take off on the intercept mission at that time (based on the Clipper's Wind Vector reports, which determined the *H6K* flight schedule)—the *H6K* Captain had to risk delay in reaching the intercept point, against abandoning the oil drums at Palau.

But the Clipper was behind schedule, and by 0011Z, he had overheard the 0000Z flight report and noted the course change toward Pos. A. Realizing that the hi-jacking was underway, although not as planned, he dared not adjust the carefully crafted timetable and took off, as scheduled, at 0016Z, with the oil drums aboard, and proceeded on the hi-jack support mission. He thus reached Pos. D, as had been originally planned, at *0300Z*, while the delayed Clipper would not reach its mirror, at Pos. D0, until *0330Z*.

Aboard the Clipper, the *Rising Sun*-Line pointers and the positions that defined them had probably been determined long before reaching Pos. A. PAA crews are known to have been deeply concerned about a possible hi-jacking, and may well have developed the principle of the LOP pointers as a means of revealing the crisis. Having determined Pos. A, Pos. B1 and Pos. B, and Pos. C and Pos. C1 (for the pointers), Davis could determine Pos. B0 and Pos. C0, and Pos. D0 and Pos. E0, which, owing to the use of Pos. A instead of Pos. A1, were offset from the original plan using Pos. D1 and Pos. E2.

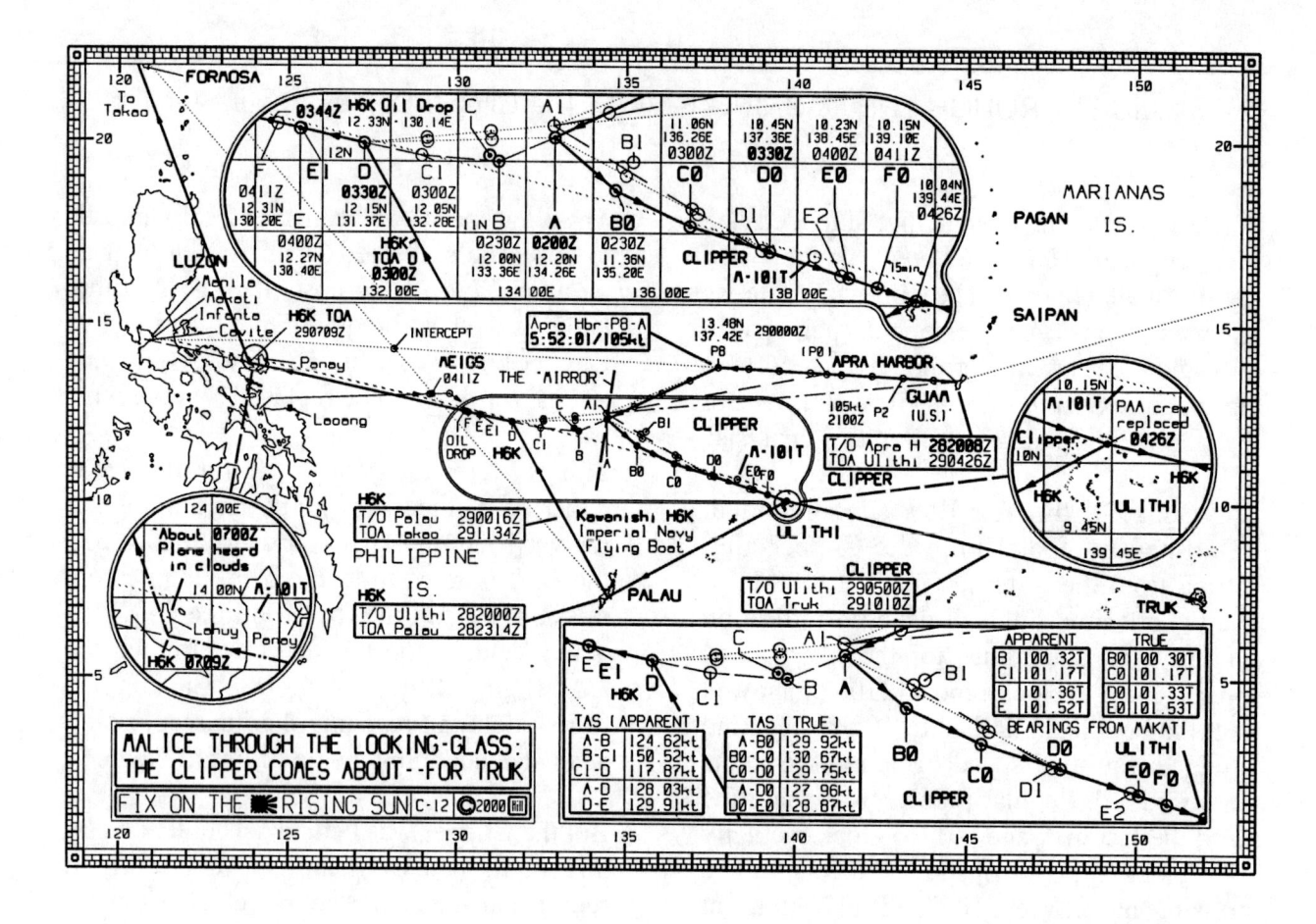

Davis had to re-route A to D0, with Pos. B0 and Pos. C0 supporting a double deception at Pos. B (B1) and Pos. C1 (C). Mirror-pairs had to meet all criteria, including an RDF match for B/B0 and C1/C0, and an increased average TAS, to provide for an *indirect* flight from A to D0, via Pos. B0 and Pos. C0, which would be equal, in flight time, to just such a *direct* flight. Davis' most difficult task may have been in convincing the hi-jackers that a detour from A to D0 was necessary, but reported severe weather may have justified the detour, as it may have justified Pos. A as an alternate to Pos. A1.

Since Pos. A was used, instead of Pos. A1, the 30 minute late Ulithi ETA was advanced by 3 minutes. That is, it occurred at 0426Z, 27 minutes—instead of 30 minutes—*behind* the original ETA of 0359Z. So, *Hawaii Clipper* reached the radio cut-off point (15 minutes from Ulithi), at 0411Z, 27 minutes after its *originally planned* "image" reached the oil slick origin site created by the *H6K* at 0344Z. Thus, the cut-off "image" occurred out of sync (in the 30 minute timing system) at Pos. F, rather than at the *H6K's* oil slick drop, 3 minutes' flight-time west of Pos. F.

With a Japanese relief crew aboard at Ulithi, the Americans had no hope of resisting, but they had succeeded in getting most of their "message" out to Panay. And, after a short layover, a flight of 5 hours and 10 minutes brought the Clipper to Truk at about 1010Z. The night landing (as also accomplished by *Philippine Clipper* at Wake Island, in 1935) served to minimize civilian observation.

The *H6K* flew on from the slick, at 0344Z, turning north to Takao over Lahuy at 0709Z, or, "about 3pm" Manila Time, as noted by the telephone company employee. But the Clipper's true "image" was not due at Lahuy until 0739Z—30 minutes later. This was one of the faults that exposed the operation.

22. 1 UNWELCOME APPROACH TO CAVITE: On Course—but a Bit too Early

On August 17, 1938, a few days prior to the official creation of the new (and supposedly non-political) Civil Aeronautics Authority (forerunner of the FAA), the very political Secretary of Commerce quickly appointed a three-man Board of Inquiry (along with two advisory members to represent the CAA's Air Safety Board, forerunner of the NTSB). The board was "to investigate the facts, conditions and circumstances surrounding the disappearance of the Hawaii Clipper, an aircraft of Pan American Airways, Inc., while flying westbound between Guam and Manila, P.I., on July 28, 1938, and to make a report thereon." The Board convened at Alameda, California, on August 18th, but immediately "deemed it inadvisable to hold a public hearing at that time..."

The Board continued its investigation in secret, and on the 22nd, President Roosevelt, by Executive Order 7959, ordered that all of the aviation files within the Department of Commerce were to be turned over to the newly formed Civil Aeronautics Authority. These included the records of the Board then investigating the Clipper and, thus, it was the two advisory members, constituting the CAA Air Safety Board, who, on November 18, 1935, released the *Preliminary Report*.

The Air Safety Board was unable to make a determination other than that contact with the flying boat was lost and never regained "and that no trace of the aircraft has since been discovered." The official press release accompanying the report explained that "the investigation remains in an open status," and the report further suggested "that additional evidence may yet be discovered and the investigation completed at that time." The report, itself, did address issues relating to the flight and to the post-flight investigation, but added a certain "spin" to its analysis:

Paragraph 8 of the *Summation* eliminated the oil slick from any further consideration, stating that "the chemical analysis of the oil sample obtained by the officers of the U.S.A.T. Meigs definitely establishes the fact that there is no connection between the oil slick discovered by them and the disappearance of the Clipper." This was not quite true, however. As already shown, there appears to be a close connection between the oil slick and the Clipper *loss*: the Board had simply added some spin to the fact that there was no *chemical* connection between the oil in the slick and the oils used on the Clipper.

Having thus diverted some attention away from the sea area, the Board continued to direct attention westward to the Philippines: Paragraph 9 of the *Summation* stated that "the report from the Island of Lahuy, that an aircraft was heard flying in that vicinity above the clouds on the afternoon of July 29, cannot be ignored." This was correct (for the wrong reasons) but aided the ASB in directing attention even further westwards: Paragraph 10 states, in part, that "the Board is not prepared to say that the aerial search of Lahuy and other areas in the islands can be considered as conclusive." The ASB then provided a last westward impetus, stating, in Paragraph 11, that a reward offered by PAA "should stimulate the search by land." While the ASB gave the impression of leaving the case "open," it was clearly predisposed to insist that the Clipper was not lost at sea.

It should be noted well, with regard to this smoke screen, that the Board recorded none of the last positions of *Hawaii Clipper*, other than the last (0400Z) position, which had already been reported widely in the press. The only source for the other positions, so far as is known, is the *Engineering Report* of PAA's Chief Engineer, Andre Priester.

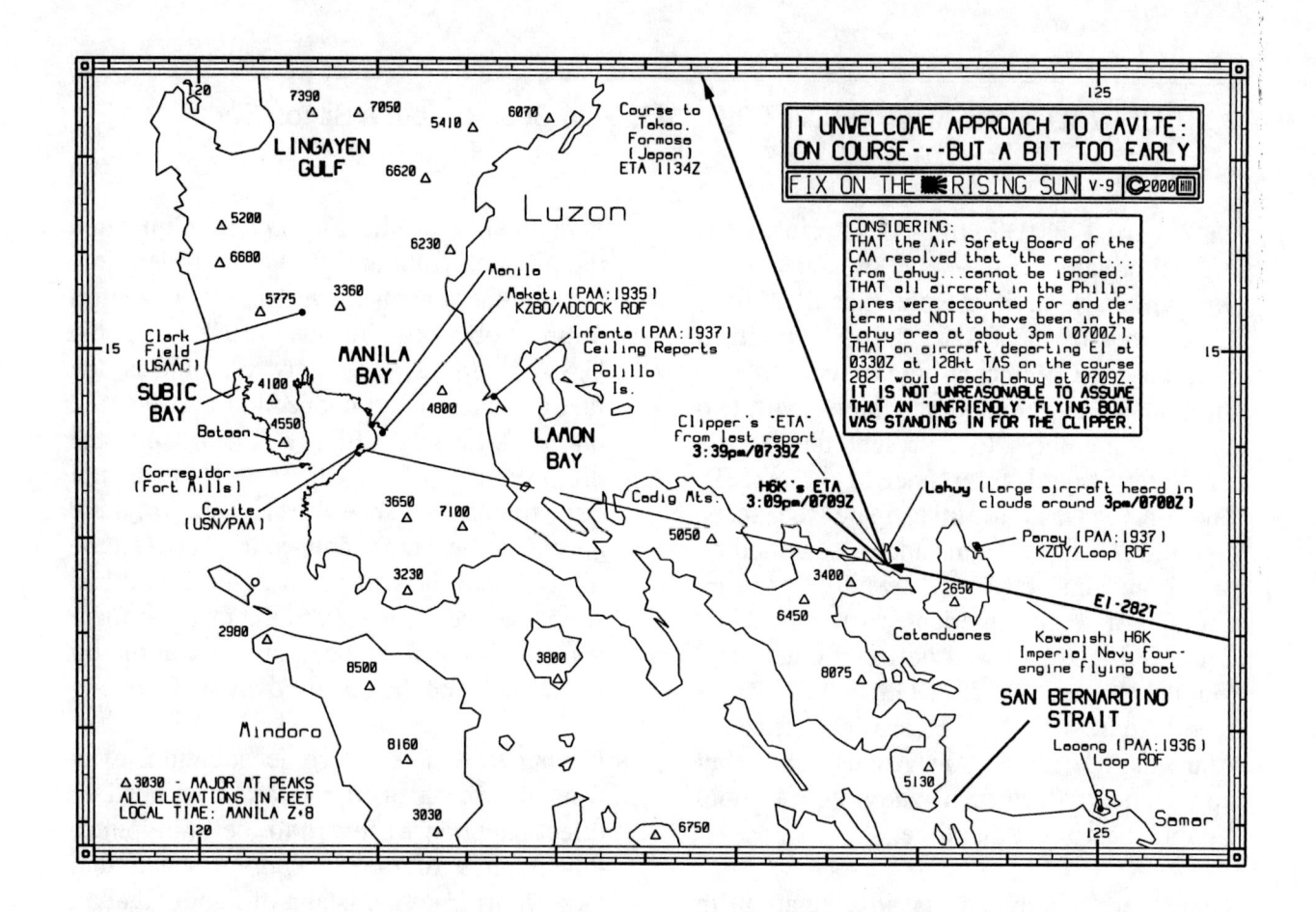

The position of the oil slick, that is, where it was located some 29 hours after the Clipper disappeared, was also missing from the ASB Report, but this important position had also appeared in the press during the search, at a time when the *Meigs*' concern with sea drift (or current set) strongly indicated that *Meigs* considered the oil slick's origin estimate to be directly related to the Clipper's loss.

As to the important Lahuy report, the Board recounted it openly, but without any critical details, noting only that "an employee of a telephone company, who lives on the Island of Lahuy, is reported to have heard a large airplane flying above the clouds about 3 P.M. Manila Time, on July 29. It was also ascertained that there were no Army, Navy or private aircraft in that vicinity on that date. Weather reports from nearby Pan American Airways Stations indicate that at the time there was an overcast, with a ceiling of about 2,500 feet, and light rain."

The ASB noted, incorrectly, that "...the Clipper should have reached that vicinity at approximately 3:46 P.M. (Manila Time)." In fact, the calculated Lahuy ETA is actually 3:39pm, some 3 hours, 28 minutes of flight time from F, or, *3.46* hours later (in decimal form). However the ASB abused it, the tantalizing Lahuy report could hardly be ignored, having already been published:

The United Press had reported, from Manila, on July 30, "The Philippine Telephone Co. advised that a reliable employee on Lahuy Island had heard a plane about three hours after Capt. Leo Terletsky of the clipper made his last [0400Z-0411Z] report," This would have been between 0700Z and 0711Z, that is, from 3:00pm to 3:11pm.

A disparity between the 0700Z sighting and the 0739Z Clipper ETA, as well as the oil slick origin estimate, make a strong case for an over-flight by an "unfriendly" aircraft.

23. DEAD-RECKONING A DEADLY GAME: The Perfect Plan—and one Minor Hitch

The Minor Hitch, of course, was an apparent delay in take-off, which led to a change in the hi-jacking schedule. The take-off delay, from 2000Z to 2008Z, was probably not sufficient, in itself, to jeopardize the hi-jack timetable, but it was part of a chain of events that, by dictating a change in fragile plans, ensured the revelation of fifteen murders.

There were no reports of excessive ship traffic in Apra Harbor, yet it seems unlikely that the delay was due to the appearance of the hi-jackers: any attempt to hi-jack *Hawaii Clipper* in Apra Harbor could have sparked escape attempts—with disastrous political results for Japan—especially if the primary hi-jacking weapon was an easily detonated high-explosive device. However, the sudden threat of obliteration, while flying at 10,000 feet, far out at sea, would have been quite effective, there being no place to run, aboard an aircraft in flight.

Moreover, reports of an unusually long take-off run, which circulated soon after *Hawaii Clipper*'s disappearance, would indicate, if true, that the Clipper was tail-heavy, which would be expected if two un-manifested hi-jackers had stowed away in the tail section.

The take-off delay was compounded by the 105kt ground speed—5kt slower than was expected. It is likely that Captain Terletsky, suspecting a possible air attack (based upon warnings then circulating In Washington), had adopted the slower speed to upset a Japanese intercept timetable. The hi-jackers, emerging from the tail an hour or so after take-off, must have found, to their bitter disappointment, that, with the delays arising out of a late take-off and the unexpectedly slow ground speed, they could never reach Pos. A1 at the required time, 0130Z, without flying at a suspiciously increased speed.

But, because the *H6K* was due to drop its oil drums at a point which was keyed to Pos. D (a false turning point), the hi-jackers had to ensure that Pos. D would be included in the Clipper flight reports. Yet this would occur only if Pos. D could be dead-reckoned at 0300Z, or even 30 minutes later, at 0330Z. Only so long as Pos. D could be calculated, as a viable position, on the hour or half-hour, could it be reported. To move Pos. D back by 30 minutes meant that Pos. A1 also had to move back by 30 minutes, to 0200Z.

Because the Clipper had lost 8 minutes in the take-off delay, the hi-jackers had only to waste 22 additional minutes en route to Pos. A1, to set their plan back "in sync." And yet, because Pos. A1 could never be reached precisely at 0200Z, while maintaining 105kt on any course to Pos. A1, Second Officer Davis plotted a precise route to Pos. A, which appeared to support the hi-jack plan. But Pos. A was offset 1 minute of arc to the east, so the *H6K* oil drop would be offset to the west of Pos. F—a revealing error.

Davis had some five hours to work out the *Rising Sun*-Line pointers, and 10 position reports by which to establish his reliability. By 0130Z, Davis had worked out all that was needed, and the hi-jackers, by then, may have come to rely on him. So, when he had to modify both routes, true and false, to provide for Pos. B1 and Pos. C (needed for the pointers), he would have been able to suggest a detour around "bad weather," and to assure the hi-jackers that he could plot this "detour" without affecting the timetable.

Undoubtedly, Davis had already calculated the detour to Pos. B0 and Pos. C0, so as to mirror Pos. B and Pos. C1, the keys to the pointers at Pos. B1 and Pos. C. To the hi-jackers, approaching the critical test of their mission, and certainly tiring under the stress of the long flight, everything must have seemed to be in order. For the next two hours, *Hawaii Clipper* flew the actual route via Pos. A, Pos. B0, Pos. C0, Pos. D0 and Pos. E0, while Second Officer Davis submitted the four false mirror positions at Pos. B, Pos. C1, Pos. D and Pos. E1 for the hi-jackers' approval. McCarty substituted a few numerical digits while transmitting the approved positions (A, B, C1, D and E1), and actually transmitted Pos. A, Pos. B1 (initially, for B), Pos. C (for C1), Pos. D and Pos. E (for E1). At Panay, Fernandez copied the positions: Pos. A (at 0200Z), Pos. B (probably B1, initially, later corrected to B, at 0230Z), Pos. C (at 0300Z), Pos. D (at 0330Z) and finally, Pos. E (at 0400Z).

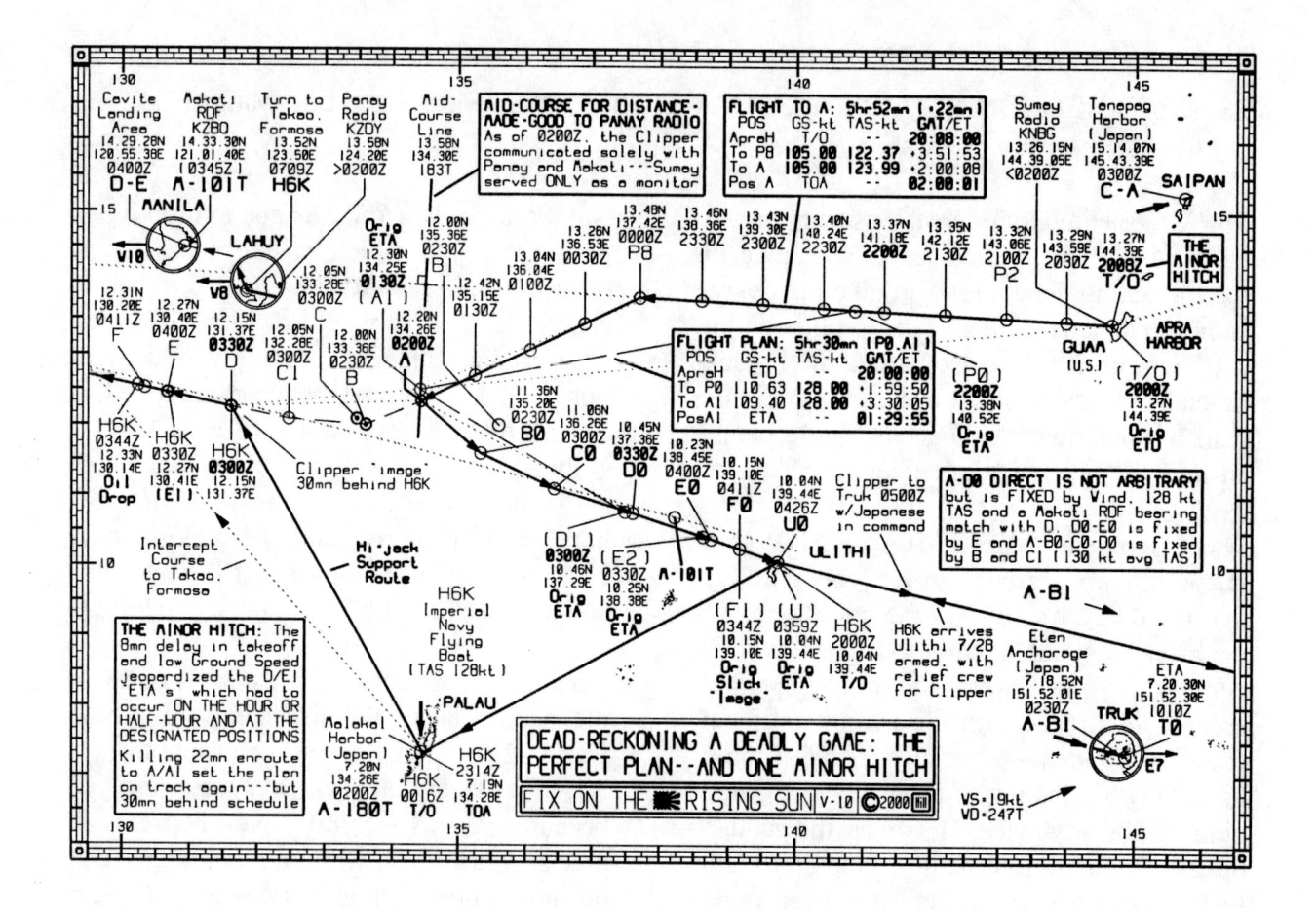

Simple enough, although it took a few years to comprehend what Davis had worked out in less than five hours. But then, he was a professional, perhaps forewarned and partly prepared, and he could not but have known, that it was the last route he would ever plot: within a day or so, he could expect to be murdered, and, for a time, forgotten. But, with a little help, he had gotten the word out.

Lives run their course, pondering "threats of Hell and hopes of Paradise." But is the end of Life so simple? Could a Soul awaken in Death, oblivious to the Life that had been so treasured? Could the butterfly forget the caterpillar—who had woven *their* cocoon?

Spirits, Ghosts, the Human Soul—are they but tales that *we* weave in Life? If a Spirit can haunt a *place*, can it not haunt a *mind*, as well? And are there not outrages so vile that a victim's Soul might reach within a *living* mind, to ensure that murder will *out*, at last?

I have come to believe, during seven years of intensive research, that there are, indeed, such outrages—and that *this* was such a one.

24. THE HI-JACK SEQUENCE: Scenario for an—almost—Perfect Crime

Even before the Clipper was declared lost, the U.S. Navy had dispatched 14 (possibly 15) search vessels of the 16th Naval District at Cavite, including three destroyers, two mine layers, a rescue vessel, a seaplane tender with two amphibians and a squadron of six submarines and its flagship.

A few days later, on August 2, the U.S. Navy announced a strange decision, which was soon outlined in the *New York Times*: "Fourteen United States Navy vessels widened their area of operations today in the fourth day of the search for the missing flying boat Hawaii Clipper...the searching craft pinned their final hopes of finding a trace of the huge Pan American Airways plane on the possibility that she may have drifted northward with the prevailing Pacific currents."

Northward? But the prevailing currents were *not* to the north! The oil slick had drifted at 1kt at 140T, to the *southeast*. And when the heavy cruiser *Indianapolis* went down, quite near Pos. A, in 1945, both her oil and her survivors had drifted to the *southwest*. If the U.S. Navy wasn't tracking a prevailing current—it was following a hunch.

On August 5, the United Press reported that the U.S. Commerce Department had been recently advised, that "Japan is considering establishment of airlines that would...furnish an airlink with the trans-Pacific Clipper service. The projected lines would run...via Guam, Takao and Palau."

The Japanese government, growing nervous, apparently hoped that the announcement of a proposed Clipper airlink would show that their hands were clean in the loss of *Hawaii Clipper*, or explain any flying boat sightings which the *Meigs* may have reported.

Around August 6, the U.S. Navy's hunch about a search to the north had apparently paid off: the U.S. State Department made the following statement, which appeared in Sunday's *NYT*: "The State Department this afternoon informed the American Embassy at Tokyo that the Pan American Airways believes the Hawaii Clipper may be in the vicinity of Parece Vela...which is in the approximate area of probable drift and would greatly appreciate a search." But the distance from F to Parece Vela Island is 575 nautical miles and, for the *Hawaii Clipper* to have drifted that far in eight days (with an additional day for the State Department to process the request) required that the Clipper drift *north-northeast* at a constant speed of 3 knots—against the prevailing current set! Sweating a bit, perhaps, Japan promised to send a *maru* to Parece Vela.

Astonishingly, the *NYT* then noted that, "it had been erroneously reported last night that the Hawaii Clipper had been found. The report, *which originated in Tokyo*, was made known in the United States by radio broadcasters and by some newspapers. *One report said that all fifteen persons aboard the plane were dead* [author's italics]."

The second lead for this unusual story was, "Airline's Request Starts False Report of Finding of Plane," but PAA's "request" (if true) could hardly have created such a "false report." The Clipper's loss was a sensitive story which had been accurately reported for a week and which, apparently, had also been accurately reported in the strange, *but true*, story which "originated in Tokyo." If any reports were false, it was the *NYT* squelch of August 6. The next morning, the *NYT* cited a Tokyo report of a typhoon at Parece Vela and an *Imperial Japanese Navy assessment*, which held little hope of finding the Clipper.

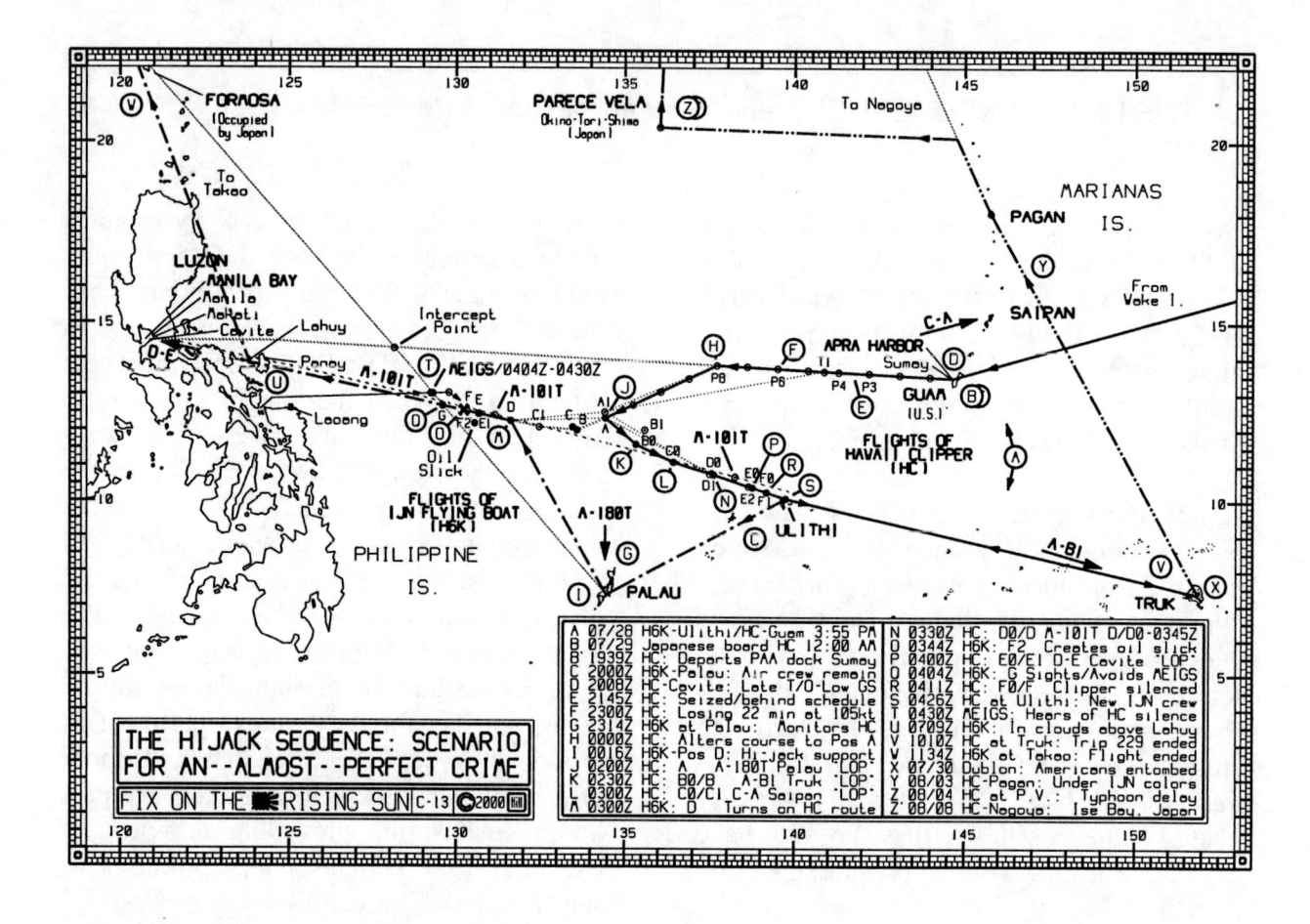

THE HIJACK SEQUENCE: SCENARIO FOR AN--ALMOST--PERFECT CRIME

FIX ON THE RISING SUN C-13 ©2000

A 07/28	H6K-Ulithi/HC-Guam 3:55 PM
B 07/29	Japanese board HC 12:00 AM
B' 1939Z	HC: Departs PAA dock Sumay
C 2000Z	H6K-Palau: Air crew remain
D 2145Z	HC-Cavite: Late T/O-Low GS
E 2300Z	HC: Seized/behind schedule
F 2300Z	HC: Losing 22 min at 105kt
G 2314Z	H6K at Palau: Monitors HC
H 0000Z	HC: Alters course to Pos A
I 0015Z	H6K-Pos D: Hi-jack support
J 0200Z	HC: A A-180T Palau "LOP"
K 0230Z	HC: B0/B A-81 Truk "LOP"
L 0300Z	HC: C0/C1 C-A Saipan "LOP"
M 0300Z	H6K: D Turns on HC route
N 0330Z	HC: D0/D A-101T D/D0-0345Z
O 0344Z	H6K: F2 Creates oil slick
P 0400Z	HC: E0/E1 D-E Cavite "LOP"
Q 0404Z	H6K: G Sights/Avoids MEIGS
R 0411Z	HC: F0/F Clipper silenced
S 0426Z	HC at Ulithi: New IJN crew
T 0430Z	MEIGS: Hears of HC silence
U 0709Z	H6K: In clouds above Lahuy
V 1010Z	HC at Truk: Trip 229 ended
W 1134Z	H6K at Takao: Flight ended
X 07/30	Dublon: Americans entombed
Y 08/03	HC-Pagan: Under IJN colors
Z' 08/04	HC at P.V.: Typhoon delay
Z 08/08	HC-Nagoya: Ise Bay, Japan

After the Clipper landed at Truk, following the hi-jacking, she soon parted company with her 15 passengers and crew. The Clipper probably spent the next few days undergoing a cosmetic makeover, with paint and insignia applied to disguise her as an Imperial Japanese Navy seaplane. She may then have been flown to Pagan to be refueled for a night flight to Ise Bay and the oblivion of a bayside hangar. But foul weather apparently forced her to divert to Parece Vela, where a U.S. Navy submarine may well have seen her, waiting out a storm.

The U.S. State Department could hardly have made such a revelation public, because the American public would have wanted blood, and the U.S. was not prepared for war with Japan in 1938. However, the U.S. State Department could reveal its knowledge of the hi-jacking and humiliate hostile elements in Japan's government and navy, simply by suggesting that the Japanese "search" for the *Hawaii Clipper* in the very place where *both* governments knew that she was stranded: Parece Vela. But Tokyo, already nervous, and suddenly faced with U.S. awareness of the hi-jacking—yet unaware that the U.S. had no intention of revealing the truth to the American public—panicked and released, to the press, the essential facts of the hi-jacking and of the fate of the passengers and crew, leaving the *New York Times* to squelch the Tokyo report, to cover for the U.S. Navy, the State Department—and the Japanese.

When the weather cleared, the Clipper was flown to Ise Bay, where she was stripped and examined by Kawanishi engineers, and where one of her big engines was copied— for the *Zero*—by other engineers, familiar with smaller Pratt & Whitney radials, but ill-equipped to design, from scratch, a twin-row engine to match those of Pratt & Whitney. After the war, a few facts emerged, but the official U.S. silence of 1938 still continues.

25. LINGERING TRACES: As Indelible—as if Cast in Concrete

Time has erased much of the evidence of the *Hawaii Clipper* hi-jacking, but some traces still remain: copies of Priester's PAA Report have been widely disseminated, preserving, despite PAA's own demise, the last reported positions of the lost Clipper, from which the incriminating *Rising Sun*-Lines are derived. Yet, material traces do remain. Throughout the Pacific, hundreds of *Zeros* rest silently, their engines bearing a simple memorial to *Hawaii Clipper*, and, assuredly, as well, to Japanese pride in the Clipper hi-jacking:

The following has been adapted from Robert Daley's 1980 biography, *An American Saga: Juan Trippe and His Pan Am Empire*. Daley heard this story from Trippe, himself, who had heard it from Admiral John H. Towers:

Following WWII, Admiral Towers joined PAA, and, after reviewing *Hawaii Clipper*'s records, informed Trippe that the Japanese had taken the Clipper. He noted that each of the early *Zero* engines had the same number cast into the magneto housing—in the place where Pratt & Whitney normally stamped a serial number. Towers told Trippe that, after examining PAA's records of magneto serial numbers for the four Clipper engines, he had found one to be identical with the number cast into the *Zero* magnetos. One can only wonder whether this was the result of haste in copying the engine or of a strange need to commemorate the hi-jacking, but, in any case, there are no surviving *Zero*'s, extant today, with the original magnetos installed. In 1992, a technician, at Wright Patterson's Air Museum, inspected the crated engine of their still un-restored *Zero* and reported that it was complete—except for the magneto.

Daley buried most details in the appendix of his book, dismissed the story in his text, and, about Admiral Towers, he provided nothing.

John Henry Towers was among the first of the U.S. Navy's aviators. He won his wings in 1911, served as assistant director of naval aviation during WWI and commanded the 1919 transatlantic flight. He commanded the carriers *Langley* and *Saratoga* and served as Deputy Commander in Chief Pacific Area during the Pacific War, succeeding Nimitz as Commander in Chief Pacific Fleet and Pacific Ocean Areas, in November of 1945. He was with Nimitz, aboard the *Missouri*, in Tokyo Bay, for the Japanese surrender, and, along with General Sutherland, MacArthur's Chief of Staff, he claimed direct knowledge of the recovery of the Clipper's engines (although a U.S. Marine reports having seen the Clipper, itself, in Yokosuka, in 1945).

Then there was the "gold." Passenger Wah Sun Choy, who owned the *China Clipper* restaurant in Jersey City, was reported to be traveling to China with three million dollars in gold, raised by Chinese-American school children, supposedly, for China war relief. It is more likely, considering the economics of the Great Depression, that these children had raised a few thousand and that Washington had sweetened the pot with a few million more, sending the money in the form of U.S. Gold Certificates, which could be replaced more easily than gold bullion, if lost in transit. If the money found at Saipan, in 1944, by Warrant Officer Horne was, indeed Mr. Choy's "war relief" fund, there may be a trace of its loss, and recovery, at Treasury.

If Wah Sun Choy's "gold" was actually a secret U.S. aid package, it was sufficient to finance fifty *Hawk 75-A* fighters (export *P-36*'s) with more than enough left over to hire American pilots and ground crews and to sustain an organized squadron in China, for at least a year. It would have created the *Flying Tigers*, three years ahead of history.

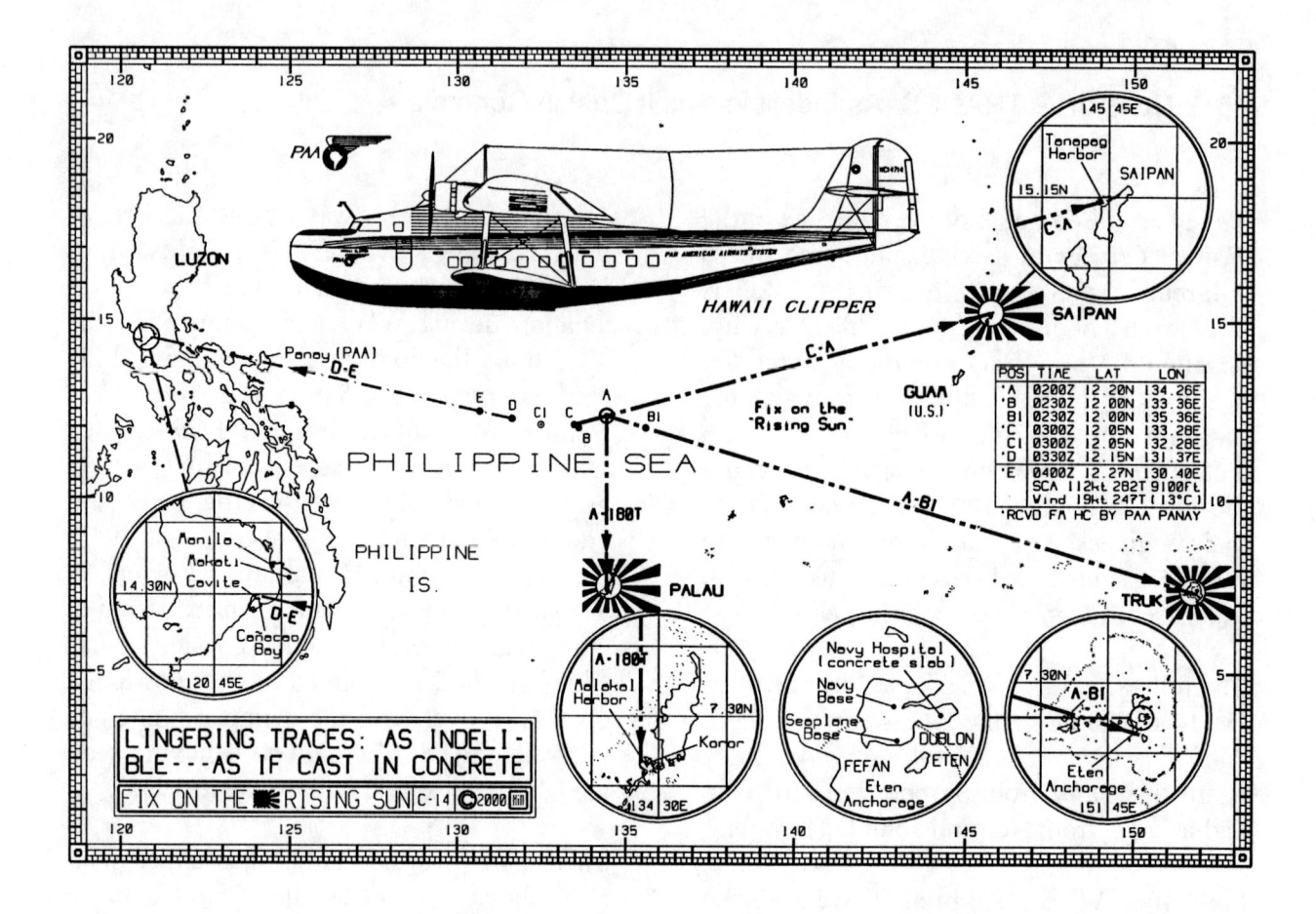

There were two additional passengers who seemed to fill the bill in this regard. Edward Wyman, a U.S. Navy Reserve Commander (and recent assistant to PAA's Juan Trippe), was a vice-president for export sales at the Curtiss Wright Company, whose *Hawk 75's* were fighting China's air war against Japan, and whose *P-40's* would later fight with the *Flying Tigers*, in 1941. Wyman's itinerary failed to detail nine curious days in "Hong Kong," but he was known to be selling engines, and China was building *Hawk 75-A* airframes, under license to Curtiss Wright.

Major Howard French, however, made no secret of the fact that he was bound for Canton, China. A successful businessman in Portland, Oregon, French had flown fighters during WWI and had served as Aeronautics Inspector for the State of Oregon. He was the commander of the 321st Observation Squadron and the senior air reserve officer in the U.S. northwest Although the *Oregon Journal* later referred to French's expensive trip as a "vacation," on July 24, prior to his departure, it noted only that it was "his first trip to the war-torn orient," and quoted him as saying, "I want to be in Canton when the Japs pull another raid on the town. I want to see how the bombs drop now." Whether or not Treasury will ever reveal traces of the recovered Gold Certificates, it is still clear that what Choy could finance, Wyman could provide—and Major French could organize.

Hawaii Clipper's hull was as advanced as her engines, and "airframe and power plant," alone, might have motivated the hi-jacking, but it was probably the threat of these three "midwives" to an unborn fighter squadron which dictated that Trip 229 be shot down, if it could not be hi-jacked. And on Dublon Island, the images of these three—and the twelve others—remain cast in concrete, even as the faces of Pompeii were preserved in the volcanic ash of Vesuvius.

26. TRUK: The Entombment Site on Dublon Island

In November of 1964, Joe Gervais, a recently retired B-29 pilot—and determined Earhart hunter—finally gained entry to Truk. He was researching a report of a twin-engine wreck, once thought to be Earhart's *Electra*. But, as is so often the case in any quest, the report proved to be in error, and the aircraft in the steamy mangrove swamp turned out to be a Japanese "*Betty*" bomber.

Gervais' guide on Dublon Island was Robert Nauroon, an elderly Franco-Micronesian who expressed a dislike of the Japanese on Truk, but who had served them as a shipping manager, and in other capacities, before the war. Despite his disliking for the Japanese, Nauroon had won their confidence, although at some cost to his popularity among the Trukese people. Nauroon had no apparent interest in Amelia Earhart, but he did have a compelling story to tell, and he insisted on telling it to Joe Gervais.

Robert Nauroon's story concerned fifteen Americans who had been brought to Truk before the war. He knew their story well, because, as he told Joe, he had helped to seal them within a slab of concrete.

In 1938, in preparation for war, the Japanese were planning to construct a small naval hospital on Unimakur Mountain, well above the swamps on the southeastern peninsula of Dublon Island. It was to be built on a poured concrete slab, 30 feet by 60 feet, oriented with the long axis along a north-south line. Nauroon, and a man named Taro Mori, were under contract to the Japanese to provide the crews to pour the slab and also to supervise the job. Mori was also a man whom the Japanese trusted, being a part-Micronesian descendant of an old Japanese family, which had established a successful trading business at Truk in the late 1800's.

According to Gervais' account of Nauroon's story, Nauroon, Mori and their crews arrived at the hospital site early one morning in the late summer of 1938. There were a number of Japanese guards and officers waiting for them, and Nauroon quickly realized why. There, in the northern half of the excavated slab site, were fifteen men, or the bodies of fifteen men, lying face down and arranged in three rows of four men each and one row of three. Some, as Nauroon reported, wore dark uniforms (as did the PAA crews), but the rest wore western civilian clothing. At some point, Nauroon was made aware of the fact that the fifteen men were Americans.

The crews worked quickly in the growing heat, and by noon, the fifteen Americans had been covered and the concrete surface was finished. Nauroon explained to Joe that they had all worked quickly because the job had turned out to be so unpleasant. But perhaps they realized, as well, that the sooner they left the site, the safer they might be: the Japanese were not above executing civilians on any pretext—and knowledge of this job would certainly have been such a pretext.

Nauroon insisted many times to Joe that the Americans had been dead when the slab was poured, but he added that, when recovered, there would be found *"no marks of death upon them."* That is, they had not been beheaded or shot, and they may well have been poisoned, as was suspected in the case of Earl Ellis, in 1923. But, according to Joe, Mori, confirming all that Nauroon had told him, had added that it had been necessary to fasten reinforcing wire over the men—"to keep them down." Mori may have been referring to a tendency of bodies to "float" in wet concrete, but he may also have been referring to men struggling to lift their heads above the concrete—as it engulfed them.

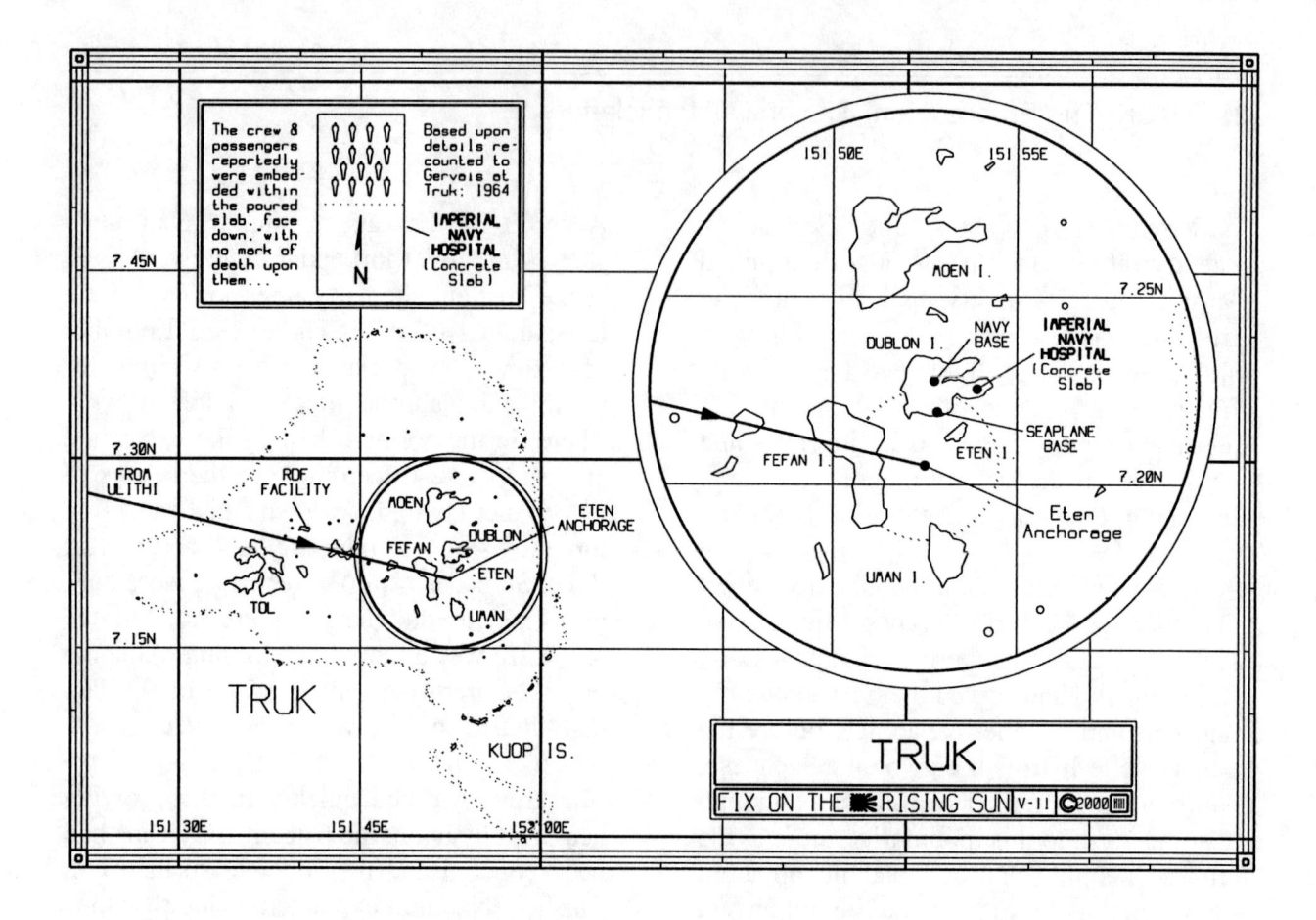

Joe was fascinated by the story, but he was then unaware of the mystery surrounding the loss of *Hawaii Clipper* in 1938 and he didn't make a connection to Nauroon's story until he read Ronald Jackson's *China Clipper*, early in 1980. Jackson's story of the *fifteen* passengers and crew of *Hawaii Clipper* set off a memory link to Nauroon's story of the fifteen Americans who had been sealed in concrete, and Joe began to reopen the case.

Joe had taken no action in 1964, because, with no awareness of the Clipper to prompt interest in Nauroon's story, it had then been just another of many Pacific tales. Then, too, in 1964, Gervais felt that he was hot on the trail of Earhart and had no desire to pursue another story. He departed from Truk, unaware that the U.S. Department of the Interior would soon sponsor a new law, forbidding private individuals to attempt to recover human remains on Truk. It is still a law at Truk (now Chuuk), even though, in 1986, the Carolines had become a sovereign nation, the Federated States of Micronesia. The law, a strange legacy of the U.S. Trust, is well known on Truk as the Amelia Earhart Law, but it now appears that its only real connection to Amelia Earhart is the bitter connection from Robert Nauroon (and the tragic story, which he most surely related to U.S. officials after the war) to Joe Gervais, the Earhart—and Clipper—hunter.

Following a disappointing attempt, in 1980, to generate public interest in the story, Joe circulated numerous copies of the PAA Report, the ASB Report and the notes of his own brief pursuit of the Clipper. It was from one of those copies, provided by fellow researcher, John Luttrell, of Atlanta, that the present analysis has been derived. In recent years, Joe has provided additional material, relating to medical war crimes committed at the naval hospital, many of them committed against American military personnel.

3. SUMMATION

1. THE EVIDENCE: A Brief Summary

1. The Rumors: One dare not discount the rumors of a pending Japanese action against the Clippers, which surfaced at the U.S. Department of Commerce in June and July of 1938, and which were documented by Ronald Jackson in *China Clipper*. This is especially true, considering the documented history of Japanese attempts to sabotage the Clippers in the U.S., as well as the F.B.I.'s report of PAA's forensic investigation of debris from the *Samoan Clipper*, in which PAA reported evidence of a high explosive having been detonated *within* the hull. With further regard to *Samoan Clipper*, the fact that Ed Musick had successfully jettisoned fuel from that type of aircraft (a Sikorsky S-42), and that the Navy had found no human remains in the debris field, surely indicates foul play, even without the forensic tests. A chilling aspect of *Samoan Clipper*'s loss was the recovery of the Flight Radio Officer's uniform jacket, with "holes blown in it."

Mark Walker's dark concern for the *Samoan Clipper*, as reported by his sister, and his belief in a possible Japanese hi-jacking, as reported by his cousin, indicate that there was genuine fear among PAA flight officers, regarding the security of the Clippers. These fears probably originated elsewhere than at Commerce, but the rumors at Commerce were certainly more widespread than was believed by Jackson. This is clearly evident in the *Meigs*' situation, astride the Japanese air route from Palau, to Formosa, at about the time of *Hawaii Clipper*'s loss, and in apparent direct communication with PAA. It seems likely that the rumors at Commerce had originated with the Army or Navy, and that Commerce had been told very little.

2. The Clipper: The *Preliminary Report* of the Air Safety Board is sufficiently detailed, as to maintenance, safety procedures, flight crew fitness and the last report from *Hawaii Clipper*, that the loss cannot be attributed to a mechanical failure or pilot error. The only problem reported (at 290411Z) was that of rain static interference with the reception of signals from PAA Radio Panay, a problem that was common enough with the exposed long-wire antenna, as used on the Clippers. Had the Clipper been destroyed by lightning at that time, as has been suggested, USAT *Meigs* would certainly have found the usual air-crash debris in the sea, during her search. In any case, for seven years prior to the loss of the *Samoan Clipper*, there had been no fatal in-flight losses of any aircraft operated directly by Pan American Airways. Between January and July, 1938, however, there had been two losses and twenty-one fatalities.

3. The Traces: Despite an extensive search of sea and land over a wide area, by some fifteen or sixteen vessels and a number of Army Air Corps aircraft, *absolutely no trace* of *Hawaii Clipper* was reported in 1938, nor has any such trace ever been reported, in any *official* release, in the nearly sixty-two years since her loss. Unofficially, however, there have been at least four traces reported: By 1946, two very senior American officers (Lt. General Sutherland and Admiral Towers) had *independently* reported the recovery of the Clipper's engines in Japan, after the war. One of the officers (Towers) noted a match between one *Hawaii Clipper* magneto serial number and one which had been *cast* into magneto housings on the early *Zero* engines. In 1964, two elderly Micronesians reported that they had supervised the burial of fifteen Americans within a concrete slab on Dublon Island, at Truk Atoll, in the summer of 1938. And, in 1997, a former Marine Intelligence sergeant informed the author that he had seen the *Hawaii Clipper* in Yokosuka, in 1945, painted as a Japanese flying boat.

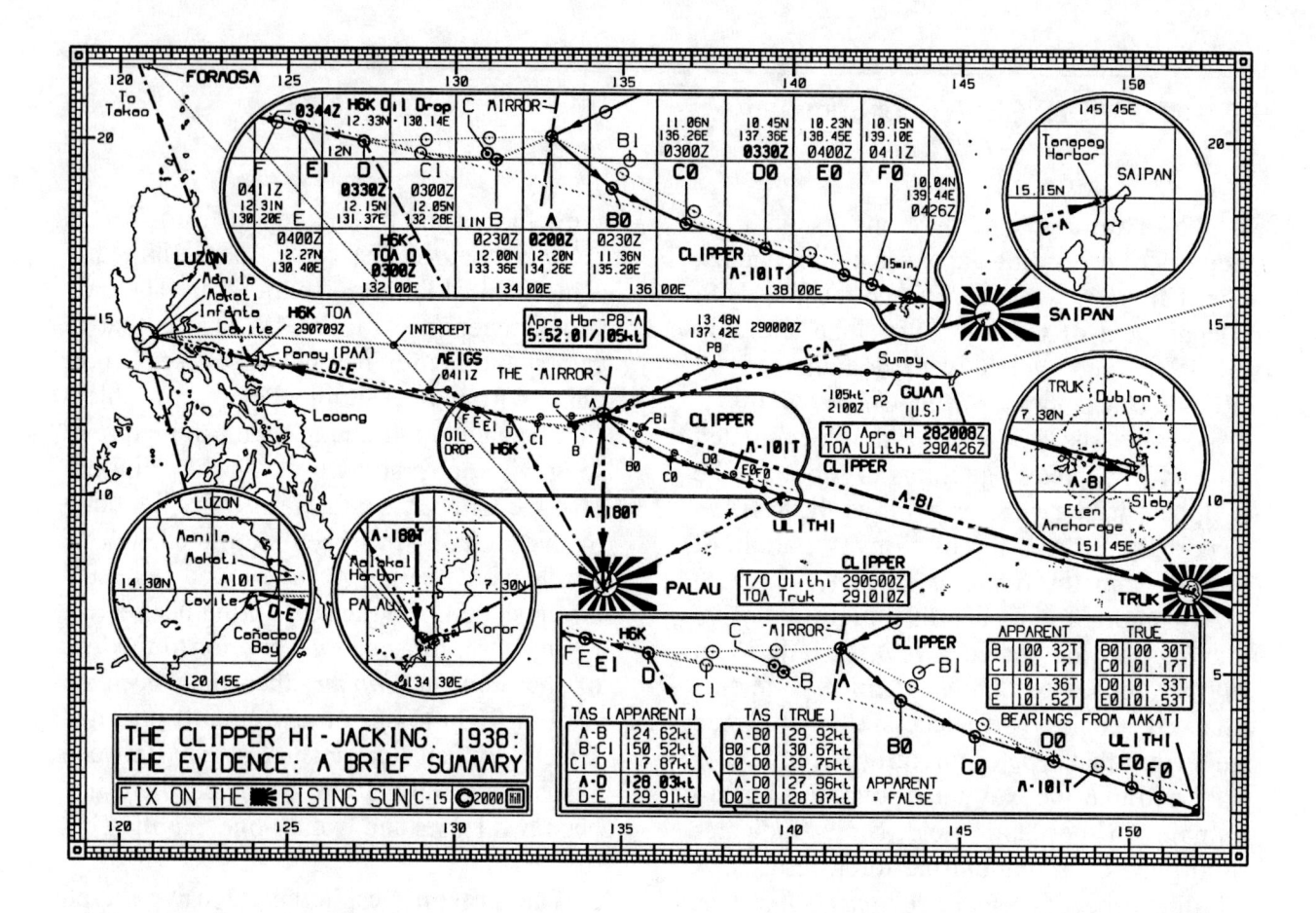

4. The Navigation Analysis: The analysis has been mathematically founded upon the five position reports received from *Hawaii Clipper* from 290200Z (12:00 noon, Guam time, +8, or K zone) through 290411Z (2:11 pm, Guam time) and recorded in the August 2, 1938 *Engineering Report* of PAA Chief Engineer, Andre Priester. Priester's report is included in the Appendix to this work. The November 18, 1938, *Preliminary Report* of the CAA's Air Safety Board provided the weather information, along with operational information relating to the performance of the Martin *M-130* flying boat. A story in the New York Times provided the only data on the early part of the flight; that is, the ground speed of 105 knots only about an hour out of Guam. All of the positions used in the math analysis have been listed in the Appendix to this work, along with the equations used for the flight analysis. The charts were created electronically and all analytical details were electronically mated to the chart. The charts were created at a scale size of 24" X 35" and would be accurate to +/- .0001," if printed at full scale. The key points in the analysis are:

A. The precise "fit" of the route from Guam to Pos. A, at 105kt ground speed.
B. The fact that, at 128kt True Air Speed, the route from A to D could be flown in 1-1/2 hours *only if flown direct*, but not, as reported, through Pos. B and Pos. C.
C. The precise crossings of landing areas by the three *Rising Sun*-Lines and the one "final course" line, the D-E LOP.
D. The mirror, at Pos. A, in which all of the false positions west of the mirror are paired with the true positions to the east, such that both positions in any pair lie along the same RDF bearing from the RDF station at Makati, preventing true positions being differentiated from false.
E. The flight integrity of the true positions, such that the true route to Ulithi is the *only possible mirror* of the false route.

5. The Oil Slick: Although the *Preliminary Report* of the ASB did not release the exact position of the oil slick, its position had been published in press articles during the search. Using drift vector (direction and speed) data provided by the *Meigs'* officers, the position of the slick, at the assumed time of loss, was easily calculated as three minutes flight time beyond the Clipper's position at 0411Z, and directly on her reported course. The ASB's rejection of the oil slick, as being unrelated to the Clipper, was made solely on the basis of the incompatibility of the oil in the slick with any oil used aboard the Clipper. Yet the Board appeared oblivious to the importance of the oil slick. It had, after all, originated on the Clipper's last reported course and, based upon the testimony of the *Meigs'* officers, it had given off an odor of gasoline, which, as was determined in the San Francisco Bay experiments, lasts only about five minutes before disappearing. The oil slick, then, was directly related, not to the Clipper, itself, but to the Clipper's *loss*. It was related by time and position (through drift) and by the fact that the lingering gasoline odor could only have come from a continuing source below the surface, perhaps even on the sea bottom. But because absolutely no debris was found in the slick, it could not have originated at the scene of a sunken air crash. It was either a weird, one-in-a-million coincidence or else a vital element of the hi-jackers' deception.

6. The Lahuy Report: As regards the report from Lahuy, the author and the ASB Report are in agreement, but for different reasons. The Board used the Lahuy report to steer further searches (and assumptions) into the jungles of mountainous southeastern Luzon. But the Board failed to compare the time of the report (about 0700Z) with the Clipper's *correctly* calculated Lahuy ETA (at 0739Z). The Lahuy report places either the Clipper, or an unidentified ("unfriendly") aircraft, at Lahuy at 0700Z, yet the timing is consistent, not with the Clipper, but with the hi-jacking support flight, as proposed by the author.

7. The Flight Reports: Although there was an apparent delay, from departure to take-off at Guam, the important factor still remains the excessively long take-off run, indicating a possible tail-heavy condition. Later press reports noted a ground speed of 105 knots, about an hour after take-off, but this was not included in the ASB *Preliminary Report*. This speed indicates, from available weather and operational data, that *Hawaii Clipper* was operating at a True Air Speed lower than its efficient 128 knots (which would have provided both maximum range and the *scheduled* twelve and a half hour flight to Cavite, at about *110* knots ground speed). The Clipper had deliberately added a half-hour to Leg 5 of Trip 229, unnecessarily and without any real explanation. However, this would have provided an unannounced delay, defensively useful, had the *Meigs* spotted a Japanese aircraft on course to the Intercept Point, between Koror and Takao. As to the position reports between 0200Z and 0411Z, it is clear that, while the Clipper had boosted her True Air Speed to 128 knots, she could only have flown from Pos. A to Pos. D at that speed and in the time allotted, if she had flown *direct*, without detouring by way of the reported positions at Pos. B and Pos. C. In addition, these latter positions represent an unlikely leg, A to B, and an *impossible* leg, C to D. Finally, the last reported course, to Cavite along the D to E line, was hardly, as Priester had noted, the "regular Southern route," but was, rather, the most hazardous route which could have been chosen (with the overcast at 2500 feet) and entirely out of character for the cautious Captain Terletsky.

8. The Passengers: Today, when an airliner is lost under doubtful circumstances, it is not unusual for the news media to inquire as to whether the presence of specific passengers might have served as a reason for a terrorist action against the aircraft. This was surely the case in August of 1938, when the CNAC *DC-2*, *Kweilin*, was shot down near Canton, en route to Chunking from Hong Kong. The aircraft was carrying the Finance Minister of the Kuomintang government of Chiang Kai-shek, as well as the head of the Chinese Air Force. It is known that China was planning to build *Hawk 75-A* airframes under license to Curtiss-Wright and that Ted Wyman, the Export Sales Vice President of Curtiss-Wright, was planning to spend nine days in Hong Kong (or China) selling engines. It is known, as well, that Jersey City restaurateur, Wah Sun Choy, was planning to hand over as much as three million dollars in U.S. Gold Certificates to the Chinese, for "war relief" (and visit his brother Frank, who was serving in the Chinese Air Force). And it is known that an Air Corps Reserve Squadron Commander, Major Howard French, was planning a nine-day visit to Canton, "to watch the bombs fall," during Japanese air attacks. It would hardly make any sense, then, *not* to assume that the Finance Minister and the head of the Air Force, having waited in vain for the rescue of these three Clipper passengers, had been targeted by Japanese naval air forces during their return flight to Chunking, thus killing (along with Wyman, French and Choy) *all* Sino-American hopes for a 1938 version of the later *Flying Tigers*. These three *Hawaii Clipper* passengers, and the mission, on which they had embarked, probably constituted a justifiable cause for Japanese action, more justifiable, perhaps, than that of the passengers aboard *Kweilin*. It is no wonder that the U.S. keeps a lid on the Clipper hi-jacking: it was not until 1992 that the government acknowledged its role in founding the American Volunteer Group, the *Flying Tigers*, who did not even fight in China, until *after* the attack on Pearl Harbor.

9. The Search: Although Priester claimed that the USAT *Meigs* had been contacted "immediately," at 0430Z, when no contact could be made with the Clipper, the Board reported that *Meigs* did not alter course to begin a search until 0740Z, at about the time (0739Z) that the Clipper should have been heard in the clouds in the Lahuy/Panay area.

Obviously, someone *had* plotted the *Hawaii Clipper*'s course from her 0400Z position and had held off the search until she failed to show up in the Lahuy/Panay area at the correctly calculated time (0739Z at Lahuy and fifteen minutes earlier near Panay). The aircraft that had been heard above the Lahuy overcast had been heard at about 0700Z, but the Clipper could barely make the speed required to have reached Lahuy by 0700Z from Pos. E. Whoever plotted her course after 0411Z would have known this, as well. It should be noted that it was *Meigs* that determined the sea drift of 140T at 1 knot when they examined the oil slick. The Navy's search is of great interest, therefore, considering that the naval sea search was later directed to the *north*, as if the Clipper could possibly have drifted against a current set to the *southeast* (140T), as measured, at the scene and at the time, by the *Meigs*.

10. The Japanese Air Route: When Japan announced plans, during the search, for an airlink between PAA Clippers via an as-yet-unlaunched Japanese airline, it should have been seen by many *New York Times* readers as an oblique disclaimer of responsibility for the Clipper loss. That is, even knowing that Japan would be suspected of a connection to the Clipper's loss, the Japanese could hardly come right out and say, "we didn't do it." However, by suggesting that they had plans for an airlink to the Clippers, it might appear that they had no reason to take action against PAA or one of its aircraft. It was clumsier than one might have expected from Japan, as if they had had little time to fabricate a more believable response, but it may have been worth the risk for a more important reason: If the *Meigs*, or any other vessel, had seen an *H6K* Imperial Japanese Navy flying boat in the assumed area of the loss, at the time of the loss, Japan would have had some real explaining to do. Mentioning a planned air route between Koror and Takao may have been all that they could do, on short notice, to explain a long-distance flight in the area.

11. The Political Games: The exchanges between Washington and Tokyo, during the search, are certainly of interest. First, if it weren't already ludicrous for the U.S. Navy to have searched for the "drifting" Clipper to the *north* of the loss area, it was well beyond bizarre for the State Department to announce that *PAA* believed that *Hawaii Clipper* had somehow turned up at the isolated Japanese island of Parece Vela (Okino-Tori-Shima), far to the *north* of the loss area, just as the Navy had strangely predicted, days before. Of course, if the Clipper were, in fact, up at Parece Vela, she could not have drifted that far, that fast, or in that direction, and PAA could not have located her. *Only the Navy could have found her*, but they could hardly have taken credit for it, as that would have stood as a territorial affront to the Imperial Navy—and the U.S. Navy was only a paper tiger in the Pacific. Yet, the Clipper seems to have been found at Parece Vela. Before the Japanese had reasonable time to send, or to pretend to send a "maru" to the atoll, Tokyo, fearing the shame of another *Panay* incident, shocked the State Department by releasing a story which, *quite truthfully*, revealed that the Clipper had been found and that the crew and passengers were all dead. But the State Department had no intention of announcing, publicly, what it apparently knew about the Clipper, as it had no options for a reprisal.

For the State Department, it was enough that the Foreign Ministry was aware that the U.S. knew just what had happened and where the Clipper had been found (as melodramatic a parting shot as the threatening, *"We know who you are, Mark Walker!"*) So, the *New York Times*, "America's newspaper," had to squelch the premature Tokyo report, which had been released by Japanese diplomats caught in yet another fiasco, not of their own making. (In December, 1937, their navy had assured them that the U.S. gunboat *Panay* had fired first on the Yangtze River, but films of the attack had proved otherwise, and had embarrassed the Foreign Ministry).

12. The Nauroon/Mori Allegations: When Robert Nauroon told his story of the fifteen Americans he had buried in concrete in the summer of 1938, it must have appeared to Joe Gervais that this was the first that any American had heard of the incident. But the murder of these fifteen Americans, and their entombment in the foundation slab of the Fourth Fleet naval hospital at Dublon Island, was not the last of the atrocities committed at the site. During the Pacific War, a number of prisoners, many of them Americans, were subjected to live medical experimentation, or, simply tortured, and finally murdered at the hospital. The end of the war came later to Truk than to the Japanese homeland, but the garrison at Truk had not held out until February of 1946, simply to keep the war going. They were busy putting fear into the hearts of Trukese Micronesians, imported Korean laborers and many others, native to the Pacific Basin. When the surrender came, the Americans were surprised to find few tales of war crimes among the former labor personnel. It took time and effort to break the stranglehold of fear, but when it broke, it broke with a vengeance. One hundred and twenty-nine Japanese were finally convicted of war crimes. Twelve of these were hanged, including an Admiral and a Surgeon. Of the forty-seven known victims, there were also *"thirteen white victims, whose nationality could not be established."* Aside from Wah Sun Choy, a Chinese-American, and Jose Sauceda, an Hispanic Texan, there were just *thirteen* "whites" among the passengers and crew of *Hawaii Clipper*. It seems unlikely that Robert Nauroon, who had no liking for the Japanese on Truk, would have told his story so freely (indeed, almost insistently) to Joe Gervais in 1964, if he had not already told it to U.S. authorities (perhaps without the gratification of official interest) at some time, or even on several occasions, in the intervening nineteen years since the war's end. [The author is grateful to Joe Gervais, who, in May of 1996, thoughtfully provided the information about the Truk war crimes.]

13. The Acts of Japanese Aggression: The fact of Japanese aggression against the U.S. in the Pacific, from 1937 through 1938, is a major element of the evidence against Japan, regarding the hi-jacking of *Hawaii Clipper*. It clearly demonstrates a disposition toward such aggression, both before and after the Clipper's disappearance.

There can be no doubt about the attack on the gunboat *Panay*, in December of 1937, at the time of the savage Japanese assault on Nanking, nor about the shoot-down of the civilian CNAC (PAA) airliner, the *Kweilin*, in August of 1938. These incidents are well documented and stand as undenied by Japan.

As to the Earhart loss, in July of 1937, the Appendix contains proof that she went down at Mili Atoll in the Marshall Islands (under circumstances that have created suspicion and interest over the past sixty-three years) with corroboration from a great number of varied sources. In short, Earhart's loss has a proven Japanese connection. The *Samoan Clipper* loss, in January of 1938, has been shown to have been, not an air crash, with all of the factors which attend such losses, but, rather, the willful destruction of a civil aircraft, on the surface, by high explosives, with absolutely no trace of its crew, except for a few articles of clothing, including a PAA jacket—the only piece of clothing that was recovered with "holes blown in it."

Of the five incidents in question (including *Hawaii Clipper*'s loss), two were undeniable and occurred in war zones, where the acts were reliably witnessed, yet one of these, the *Panay* incident, an unofficial action by a few renegades, was blatantly misrepresented by the Japanese Navy to their own government. Of the other three, all involved the loss of an aircraft and crew (and passengers on *Hawaii Clipper*) and appear to have been elements of the on-going progressive development of an increasingly sophisticated hi-jack system. And two of these have connections to Japan.

14. The *H6K* Support Flight: Admittedly, of all the elements of this body of evidence, the presence of a Japanese aircraft, that is, of a four-engine Kawanishi *H6K* flying boat, operating in support of the Clipper hi-jack mission (with an alternate attack mission), may appear to be more speculative than can be supported by the evidence, especially as it has no discernible *material* existence. But it left a trail, like a predator in the bush.

The news from Japan, proposing an airlink to the Clippers, while serving as a means of diverting suspicion from Japan, was just too clumsy and obvious not to have had another purpose, that of explaining the presence of a Japanese flying boat in the hi-jacking area, had one been seen by the *Meigs* or, through a break in the clouds over Panay or Lahuy.

The presence of the *Meigs*, along the track from Koror to Takao (one of the proposed Japanese air routes), indicates that the route had been suspected of being, or was known to be, the cover for the impending attack on *Hawaii Clipper*, as suggested by the limited rumors which had circulated at Commerce.

A flight along this track from Koror, at the take-off time suggested, not only provided a properly timed drop for the gas-oil source (at the proper place), but also met the later requirement for a pass over Lahuy, within nine minutes of the dead-reckoned time.

The oil slick *existed*, had a limited duration, was fed by a continuing underwater source (as proven by the rapid evaporation of gasoline observed in the *Meigs*' static tests), had to have been delivered to the site by some means and was delivered at a time and place which meets all requirements for an association with the Clipper. The "sighting" at Lahuy *occurred*, and, having occurred too soon to have been *Hawaii Clipper*, and there being no "friendly" aircraft in the skies, in that area, at that time, the overflight by an *H6K* is not to be dismissed lightly.

2. THE SCENARIO: Tying the Evidence Together

The subtitle of Chart C-0 defines it, and, in fact, this entire work, as an "Indictment for Piracy and Murder." In cases before a grand jury, where the evidence is circumstantial, or where the crime is particularly complicated, it is customary for the prosecutor to tie the evidence together by providing a scenario, based upon the evidence. Such scenarios not only outline the events of an alleged crime, but also provide a grand jury with insight as to the prosecutor's perception of the events, as supported by the evidence. A grand jury's task is not to determine innocence or guilt, but only whether a prosecutor's perception, as illustrated by the evidential scenario, is sufficient to indicate, first, that a crime has been committed and second, that allegations against an accused warrant the scrutiny of a trial, under the tighter rules of due process.

A. 7/28/38 3:55pm, Thursday: Ulithi/Guam

At Ulithi, landing after a flight from Truk, an Imperial Navy *H6K*-type flying boat, Kawanishi's improved copy of the Sikorsky *S-42*, taxis to a small island and moors at the Japanese weather radio station. On board are the regular flight crew, as well as a special relief flight crew, who are to remain at the station for the night. Far to the northwest, at Apra Harbor, Guam, arriving from Wake on Leg 4 of Trip 229, the *Hawaii Clipper* lands, taxis to the PAA terminal slip at the former Marine base at Sumay, and, after her crew and passengers have disembarked, is fitted with an undercarriage. After ground crews have winched her up the ramp, she is towed to the PAA terminal hangar for overnight security and inspection. Unnoticed in all the activity, two men, Imperial Japanese Navy officers, both well disguised as Chamorran laborers and briefed by Japanese espionage agents, slip into the hangar and settle into the noisy refrigeration compressor room.

B. 7/29/38 12:00am Friday: Sumay, Guam

Sometime after midnight, with maintenance completed and watchmen on alert outside of the hangar, the two Japanese emerge from hiding and board the Clipper through one of the two emergency doors which open onto the sponsons. Once inside, they proceed to the tail storage compartment and begin to repack the contents of several thoughtfully designed crates. The crates are the work of Japanese operatives who have arranged to ship them through a non-Japanese agent in San Francisco. Working swiftly, the two hijackers, now wearing PAA flight uniforms, establish sufficient space, within two crates, both for themselves and for their equipment, which includes handguns and a vest-bomb, and settle in to await the morning take-off.

B'. 281939Z 5:39am, Friday: Sumay, Guam

In the early morning twilight, under constant observation by PAA ground crews and flight personnel, *Hawaii Clipper* has been towed out to the seaplane ramp, launched into the terminal slip, relieved of the burden of its temporary undercarriage and prepared for boarding. With passengers and crew aboard, the Clipper is towed out into Apra Harbor for take-off. Assistant Engineer Officer Tom Tatum starts the engines and Steward Ivan Parker conducts a routine preflight tour, which reveals nothing unusual, even though he does pay particular attention to the tail storage compartment. The equipment crates, having already been loaded on board and examined in California, are of little concern during this inspection. With six passengers on board and nine crew to fly and train aboard her, *Hawaii Clipper* then waits at the harbor's margin, while a PAA motor launch drags the take-off area, searching for debris and signaling harbor traffic to stand clear.

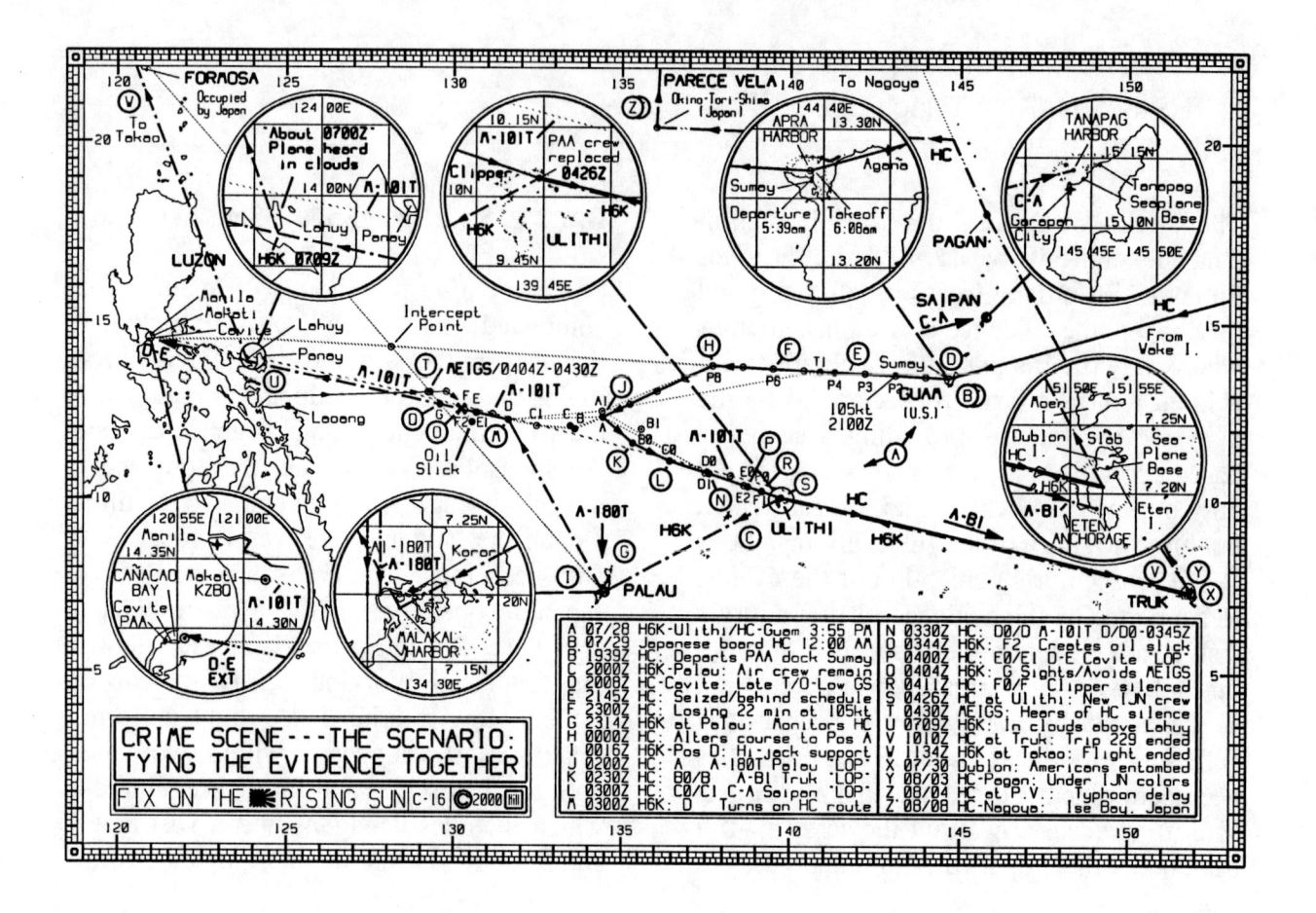

C. 282000Z 6:00am: Ulithi

In the warm predawn at Ulithi, the relief flight crew stand by with the radio station personnel and watch, as the Imperial Navy *H6K* slips her moorings and sails slowly away from the small island and out into the dark, choppy waters of the open lagoon. The Captain and his co-pilot lose themselves in the take-off ritual, ticking off the checklist items and noting coral heads in the lagoon. Then, as the *H6K* wallows, surges, bounces and finally lifts off, the navigator notes the time, for precise navigation on their rigid flight to Palau, and the radio operator sets his receiver, to listen for reports of *Hawaii Clipper*'s departure. In the cabin, a crewman stares at an unmounted machine gun.

D. 282008Z 6:08am: Apra Harbor, Guam

With John Jewett, in training, at the controls, *Hawaii Clipper* sails clumsily into Apra Harbor. She wanders, as she waits for an "all clear" from the launch, and when it comes, she's out of position for take-off. When she finally makes her run, she seems almost unwilling to rise out of the harbor, bouncing over the surface for twenty seconds longer than usual, before taking to the air at Orote Point. Jewett reports that she felt tail heavy, but he is in training and his observation is noted with marginal interest. Second Officer George Davis, as part of his regular duties, will check the after compartments during his mandatory post-take-off inspection tour, but he will see nothing out of the ordinary.

In the air and coming to course 273T, nearly direct to Cavite, Captain Terletsky orders that *Hawaii Clipper* be flown at 105 knots ground speed, not the planned 110 knots, as a precaution against a recently rumored, and long expected, Japanese naval air assault on one of the Clippers. Terletsky, who usually spent time talking with his passengers, had realized that three, alone, on this trip, could be considered cause for a Japanese attack.

But the low ground speed is not Terletsky's idea, alone. It is a part of a far larger plan, involving the U.S. armed forces: A friendly ship, the U.S. Army Transport *Meigs*, lies along a direct course from Koror, in Palau, to Takao, in Formosa, well to the south of the point of intersection with the Clipper's route. If a Japanese flying boat is seen by the *Meigs*, at or about 0349Z, then a probable intercept will be apparent and the Clipper, actually flying behind its planned schedule, will be able to take evasive action well in advance of the intercept point. As to other dangers, however, although the potential for a hi-jacking has been foreseen by some PAA officers, PAA management has ordered no special preparations for such an outrageous, and, in some minds, unlikely, action.

E. 282130Z 7:30am: 13.36N/141.45E (P3)

About an hour and a half into the flight, at a time when the passengers and the off-duty crew can be expected to have settled into the languor of a fifth long day of Pacific flying, the two Japanese hi-jackers arm themselves, leave the tail and move quickly to secure the lounge and the passengers' compartments. The lead hi-jacker graphically defines, for the startled Americans, the function of the suicide vest-bomb, as worn by his partner. The pair then move forward to secure the bridge and its crew and to disable the radio, at least temporarily. They intend to order a course change at 2200Z, only minutes later, to Latitude 12.30N, Longitude 134.25E but are surprised to find that the Clipper is far behind her usual schedule. She is unable, at normal speed, to reach the critical mirror-point [A1] at precisely 0130Z, when it must be reported to Radio Panay. With his critical plans in disarray, the lead hi-jacker disables all of the radio transmitters, one of them only temporarily, and orders the Clipper to proceed on course at her present speed. With his control of *Hawaii Clipper* assured, but death awaiting their arrival at the intercept point, just a few hours ahead, he commits himself, fully, to the seemingly impossible task of recovering the original hi-jack plan.

F. 282300Z 9:00am: 13.43N/139.30E (P6)

Once in full control, with the passengers and extraneous crew restrained and his partner, the bomber, occupying the engineer's seat in the cabane (from where he can observe the bridge), the leader supervises two uneventful standard radio exchanges. In addition, with reluctant, but skilled "cooperation" from the Navigator, Second Officer George Davis, he initiates a new plan, which requires that they kill some 22 minutes while cruising to a new mirror-point at the initial 105 knot ground speed. They will remain on the Direct Route until 0000Z [P8] and then turn toward a new mirror-point [A], arriving at this position at 0200Z, 30 minutes behind schedule, but back "in sync" with the original plan. There, they will increase the True Air Speed to 128 knots and turn toward Ulithi Atoll, on a new route which, to PAA's Makati RDF, will be indistinguishable from the route to Cavite.

G. 282314Z 9:14am: Koror, Palau

Landing to refuel for his contrived "survey flight" to Takao, Formosa, the *H6K* Captain is puzzled as to which of the two missions he must fly. If *Hawaii Clipper* had been hi-jacked, she would have changed course for the mirror-point at 2200Z, and the *H6K* would leave Palau at 0016Z, on the hi-jack support mission. But, with no hi-jacking, he has to take off no later than 0000Z, in order to intercept the Clipper after 0439Z, on his course to Takao, and, either force her down, to achieve the hi-jacking, or else destroy her in the air. But he sees that, although she has made no course change, her ground speed is low, which will put her behind schedule to the intercept point. And so, the *H6K* Captain takes a risk: he orders the loading of two oil drums, modified for the hi-jack support, and waits for the Clipper's 0000Z report.

H. 290000Z 10:00am: 13.48N/137.42E (P8)

On *Hawaii Clipper*'s flight "bridge," the operations appear almost normal. The Third Officer, Jose Sauceda, now in the left seat, flying as pilot, completes the turn to 245T, assisted by Fourth Officer Jewett, who sits to his right as copilot. Flight Radio Officer McCarty, facing forward in the tight radio position behind Jewett, waits as, just above and behind him, Second Officer Davis, also facing forward at his chart table, completes the 0000Z report. But behind Davis, in the Engineer's position, high in the "cabane," the bomber watches intently as his leader glances at the report and then hands it, along with a telegraph key, to FRO McCarty. Only perfunctorily supervised, McCarty plugs in the key and begins transmitting to Sumay. Davis notes that they are gaining the hi-jacker's confidence, even though it is only the contemptuous confidence, common to all terrorists, who seem to relish the humiliating "cooperation" which is always born of their victims' dark fears and falser hopes. Davis, however, is in no way humiliated.

I. 290016Z 10:16am: Koror, Palau

At 0007Z, monitoring the Clipper's report, the *H6K* Captain realizes that his risk was justified and that the Clipper had been truly hi-jacked. She had changed course at 0000Z, for the mirror-turn to Ulithi, but, at 0011Z, as the report ends, his Navigator sows a seed of doubt: she is two degrees off course to the planned mirror-point [A1], and her ETA is 0158Z, twenty-eight minutes off schedule. Suspecting a possible counter-deception and a sudden, silent sprint to Cavite, the *H6K* Captain anticipates a mission change and retains the machine guns, already aboard. He departs on schedule, as planned, at 0000Z, staying ahead, that is, west, of the Clipper, as he begins the hi-jack support mission.

J. 290200Z 12:00pm: 12.20N/134.26E (A)

As the Clipper reaches the mirror-point, Radio Sumay prepares to assume the duties of monitor and transfers all communication responsibility, for Trip 229, to Radio Panay and the Makati RDF. At 0203Z, the Clipper reports turning onto course 248T, en route to Cavite, but actually turns onto course 130T, en route to Ulithi. To the RDF at Makati, all the subsequent position-pairs, true and false, will be mirror images along the same RDF bearing, and there will be no way for the single RDF (at Makati) to tell false from true. With all operations normal, Sumay will act as a monitor and take no bearings unless requested to do so. Moreover, the lead hi-jacker has had the transmitter power reduced so that Sumay will receive few signals. The lead hi-jacker is pleased that Davis is so cooperative, but, in his contempt for what he perceives as fear-driven cooperation, he fails to note that Davis' new mirror-point [A] lies precisely due north of Malakal Harbor.

K. 290230Z 12:30pm: 11.36N/135.20E (B0)

Davis, up to now, has done his work alone. He has gained the confidence of the lead hi-jacker with his revised route and its new mirror-point. He has suggested that, on a "weather related" southern detour, dodging of squalls is typical, so, now they are flying Davis' deceptive route "around the squalls" and setting up his plan "to get the word out." Now, Davis must depend upon McCarty to substitute, as he transmits, *one* amended or *reconfigured* digit of longitude for a specific written digit, approved by the lead hi-jacker. Davis cautiously shows McCarty his means of indicating changes, and McCarty sends the first substitution, which defines a line from the new mirror-point [A], through the substitute [B1 for B], and on, precisely, to Eten Anchorage at Truk. However, Panay requests a repeat, and McCarty, fearing that he may arouse suspicion and expose Davis' plan, sends only the approved position [B].

L. 290300Z 1:00pm: 11.06N/136.26E (C0)

Undaunted, Davis passes McCarty another substitution at 0300Z, defining a line from the substitute [C for C1], to the turning point [A], and on to Tanapag Harbor, Saipan. This substitute, not quite so far displaced as to be immediately questionable, is copied, without comment, by Edouard Fernandez, at Radio Panay, despite its technical unsuitability for the Clipper's true position. Although he has lost the substitute [B1 for B] for the *Rising Sun*-Line to Truk, George Davis has, so far, defined two lines through the mirror-point, each one of them fixed on an Imperial Navy seaplane base. With one substitution left to send [E for E1, at 0400Z] Davis has hopes that, once Pos. E is received, someone at PAA will tumble to his plan and eventually dope out the missing substitute [B1 for B]. With all three *Rising Sun*-Lines charted, PAA will have a "*Fix on the Rising Sun.*"

M. 290300Z 1:00pm: 12.15N/131.37E (D)

Assured by his own RDF loop, that the Clipper is en route to Ulithi under Japanese control, the *H6K* Captain turns onto the Clipper's false course to Cavite, still on the pre-planned hi-jacking schedule, but a full thirty minutes ahead of *Hawaii Clipper*'s reported progress. Aft, in the *H6K*'s cabin, the two oil drums are moved close to an outer hatch and the protective top covers are removed. At 0344Z, in support of the hi-jacking, they will be dropped into the sea, along the false route.

N. 290330Z 1:30pm: 10.45N/137.36E (D0)

As the Clipper actually turns onto course 108T, to Ulithi, McCarty begins transmitting the false report, showing a turn onto course 282T, to Cavite. At 0345Z, after sending the report to Panay, McCarty keys a series of "M.O.'s" to the Makati RDF, which reports a bearing (QTE) of 101T. This appears to confirm the reported position [at D], even though, at 0330Z, the Clipper was actually some 363 nautical miles east-southeast [at D0], and heading in the opposite direction. But, so long as things appear normal, Sumay will take no bearing, which, plotted against the Makati bearing, would fix the Clipper's true position. The lead hi-jacker will soon order a visual sighting on Ulithi, ostensibly for a continued flight to Eten Anchorage at Truk. He believes that the Americans will remain calm, hoping for an opportunity to resist, between Ulithi and Truk, but he will not risk such a long flight and, at the last moment, he will order a landing at Ulithi.

O. 290344Z 1:44pm: 12.33N/130.14E (F2)

Keeping to his tight, well planned schedule, but thirty minutes ahead of the Clipper's false route, the *H6K* Captain drops to a few hundred feet altitude and orders the two oil drums cast into the sea. Weighted at the bottom and not quite full, the drums will sink upright until, at a depth of ninety feet, small scored areas on the drum tops will implode and allow the gasoline/oil mix to seep out, thus creating a lasting, realistic "air crash" surface slick. This position [F2] is the mirror of the planned position [F1], at which the Clipper was originally expected to be on her final approach, just fifteen minutes out of Ulithi. At this time, 0344Z, her radio was to be silenced, after regular communication schedules were completed, and the oil slick from the drums would confirm a false "crash site" [F2], west of the mirror. However, because the mirror-point was shifted [from A1 to A], the Clipper will reach its approach point to Ulithi [F0], not at a precise thirty minutes behind the hi-jacking schedule, but at twenty-seven minutes behind, at 0411Z, or, three minutes ahead of "sync." Thus, the oil slick, dropped at a pre-planned position, will actually appear to have formed three minutes "flying" beyond the mathematically precise false mirror-"crash site" [at F]. It is this disparity, above all, which puts the two flights "out of sync" with each other.

P. 290400Z 2:00pm: 10.23N/138.45E (E0)

Aboard the Clipper, Davis advises McCarty of his final substitution. The false position [E1] is most proper for navigation, in that it best expresses Distance Made Good along the course to Cavite, but the substitute [E] defines a course [D to E], which will extend, precisely, to the Cavite landing area. Davis intends that this Line of Position serve as a key to recognizing the previous lines and believes that it will underscore the obvious fraud of such a dangerous foul-weather route to Cavite. After a quick crew exchange, the other Americans are made aware of Davis' work and realize that, even if no opportunity for resistance opens up, en route to Truk, there remains, at least, the chance of rescue.

Q. 290404Z 2:04pm: 12.41N/129.36E (G)

Aboard the *H6K*, the Captain decides to turn north on his scheduled "survey" route to Takao, aborting the optional "foul-weather" flight to Luzon. Such a flight, if briefly seen or heard, could divide the search for the Clipper, but by making up a little time on the route to Takao, he can obscure his own part in the hi-jacking, should anything else go wrong with the plan. Before he makes the turn, his Co-pilot sees a faint ship's wake, some thirty miles to the northwest and then, almost directly on the Takao route, the ship, itself, the *Meigs*, pausing en route to Guam to guard against a Clipper interception. The *H6K* remains on the false Clipper route and, at 0416Z, risks being sighted, as it passes within sixteen miles of the *Meigs*.

R. 290411Z 2:11pm: 10.15N/139.10E (F0)

At 0411Z, as McCarty finally completes his transmission of the 0400Z report, the lead hi-jacker notes that the Clipper is just fifteen minutes from Ulithi. Edouard Fernandez, at Panay, then requests that McCarty copy a weather report, but the lead hi-jacker orders McCarty to reply in Morse, "AS1 QRN," or,

"Wait one minute. I am experiencing static." The hi-jacker then demands the return of the radiotelegraph key and orders McCarty into the mail compartment, directly beneath the bridge. Settling into McCarty's now-vacant position behind the co-pilot, with McCarty in view below and his accomplice watching Davis from behind, he orders Terletsky and Walker, flying as Pilot and Co-pilot, to drop through the overcast, so that Davis can take his sighting on Ulithi. Terletsky brings the Clipper down slowly, as Engineer Officer Cox, in the cabane, works gingerly near the Japanese "bomber," reducing engine power. As they drop out through the overcast, only two thousand feet above the sea, all eyes are on the horizon, looking for signs of the large atoll. Within minutes, Davis sights it, and the hi-jacker calmly suggests a fly-over.

S. 290426Z 2:26pm: Ulithi Atoll

The lead hi-jacker announces that the men at the radio station will be surprised to see the Clipper and points toward a small island, suggesting that they fly over it. As Terletsky brings the Clipper around to over-fly the island, the lead hi-jacker turns and reminds Davis to take a fix on the island, for their long flight to Truk. With Davis occupied, he draws his pistol, takes a last look around to make sure that everything is in order and orders Terletsky to land in the seaplane lane near the island. Terletsky hesitates, but when the lead hi-jacker repeats the order, he sets up a final approach to the landing area. In the lounge, the Americans' spirits fall, as they realize that they are about to make an "unscheduled" stop at Ulithi.

After touchdown at Ulithi, Terletsky sails his Clipper into the wind for the last time, waiting for the station motor launch and the Japanese relief crew, who are to fly *Hawaii Clipper* on to Truk. The other Americans are especially dismayed, for they had all truly believed that fatigue and stress would take a toll upon the hi-jackers, en route to Truk, and open a window of opportunity for some resistance. The lead hi-jacker shares their dismay, for he had believed that he could take the Clipper to Truk, *without* any relief crew, thus reserving all of the "glory" for himself. However, he realized that stress and fatigue would have weighed heavily on his partner, the bomber, and, of course, that was the weakness against which the Americans might have prevailed. At about 0500Z, with all prisoners restrained, the relief flight crew prepares to take off for Truk. The lead hi-jacker remains, technically, in command of the Clipper, but, like Terletsky, he will leave most of the flying to the relief pilots.

T. 290430Z 2:30pm: 13.01N/129.18E

Aboard USAT *Meigs*, the bridge officers are surprised to learn, from KZBQ Makati, that all communication with *Hawaii Clipper* has been lost and that she had reported severe weather at 0400Z. Advised of her reported course, just to the south, the *Meigs*' officers, who have experienced no severe weather in the area, believe that a possible sighting, just minutes before, may have been the Clipper, flying along her reported route, apparently on course for Cavite, but, quite possibly, without a working radio transmitter.

U. 290709Z 5:09pm 13.53N/123.50E P.I.

Flying above the overcast and confirming his position by monitoring Panay's calls to *Hawaii Clipper*, the *H6K* Captain reaches the point where he must turn north to Takao, to avoid the risk of flying over northeastern Luzon. Turning away from the false Clipper course, thirty minutes ahead of the Clipper's "image," he passes over Lahuy Island, above the overcast. An employee of the Philippine Telephone Company hears the four engines of the *H6K*, and, although he is unable to see the aircraft through the overcast, notes the time as being "about 3:00pm" (Manila time; 0700Z). This "sighting" will serve to divide the search for the Clipper, but no Navy search will begin until she is well overdue at Cavite, at 0911Z. The U.S. Navy believes, for several critical hours, that she may still be flying on course, but without a radio.

V. 291010Z 8:10pm: Eten Anchorage, Truk

Pan American Airways' Trip 229 terminates at Eten Anchorage, Truk Atoll. There is no fanfare as *Hawaii Clipper* sails slowly up to the mooring at the Dublon Island seaplane base. All who are present understand that the Imperial Japanese Navy has committed an act of piracy, on and over the high seas, and the enormity of what has been done rests uneasily on the few who watch the Clipper approach in the evening twilight. The true import of the crime is grimly delineated as the prisoners are brought ashore, for no one, not even among the Americans, is oblivious to the fact that the act of piracy must soon be compounded by fifteen acts of murder.

W. 291134Z 9:34pm: Takao, Formosa

After her long flight, the *H6K* touches down at Takao, on Formosa's southwestern coast. She is just over an hour behind schedule for the "survey flight" from Palau, but there will be little rest for the weary crew. Dawn will find them en route to Tokyo, to uphold the ruse of a "route survey" for commercial air service from the Mandates in the western Pacific. In the many hectic days ahead, and desperate to explain any possible sightings of Japanese flying boat activity down in the Philippine Sea, Japan will release the details of these tentative "air routes," daring even to suggest an improbable airlink, *at Guam*, to PAA's trans-Pacific Clipper route.

X. 08/01/38 Monday: Dublon I., Truk

On Saturday morning, the fifteen Americans are taken by truck to Unimakur Mountain, on the southeast peninsula of Dublon Island. At a cleared site, overlooking Eten Island, they are told that they are to be put to work, helping to build a new naval hospital, until arrangements can be made to repatriate them to America. Before beginning work, they are offered a simple breakfast, which includes, to their surprise, hot coffee, fresh from the Clipper's galley. A few of the Americans are reluctant to accept food or drink from the Japanese, but after seeing the swift effect of the poisoned coffee on their compatriots, they elect to follow this path, rather than to challenge their Japanese captors to employ a more violent means of murder. For a few of the Japanese officers and military police at the site, there is an unspoken relief that the fifteen Americans have committed "suicide" and that Japanese hands have thus remained "clean," regarding their deaths. The military police then arrange the fifteen bodies, face down, in the excavation, in three rows of four men each and one row of three each.

When the contractors arrive with their local crews, they are surprised—and uneasy—to find that the northern-most of two thirty-foot by thirty-foot shallow excavations contains the bodies of fifteen men—several wearing dark uniforms. The contractors are told that the fifteen men are all Americans, and, after mixing the concrete, the local crews work swiftly to fill the excavation. The first slab, containing the Americans, is completed by noon, and the second is nearing completion in the late afternoon, when the laborers are offered a cool drink by the Japanese, as a reward for their efforts. In a very short time, the two contractors, who are well known to the Japanese and trusted for their silence, finish the second slab—alone.

With the Americans murdered, the Japanese government cannot possibly right the wrong, yet, the bodies still remain available to the Fourth Fleet, as a means to blackmail their own government, should Tokyo ever oppose the Fleet Faction radicals at Truk. With the laborers dead, there will be no gossip, and there is such fear of the Japanese at Truk that these missing men will never be sought.

Y. 08/03/38: Wednesday Truk-Pagan

Under new colors and painted as a flying boat of the Imperial Japanese Navy, *Hawaii Clipper* takes off late Wednesday evening for a night flight to Pagan, in the Marianas, en route to Ise Bay and Nagoya, in Japan. The lead hi-jacker, now properly uniformed, is again in command of the Clipper and most eager to present his prize in Japan: the new fighter under development, the *Zero*, will fly with a powerful fourteen-cylinder twin-row engine, developed from the Clipper engines, engines which he is bringing to Japan.

Z. 08/04/38: Thursday Pagan-Parece Vela

Refueled at Pagan, but with a severe storm to the north threatening the flight to Japan, the lead hi-jacker decides that the risk of weather is less than the risk of an American over-flight from Guam and takes off for Nagoya. Late in the morning, with the storm clearly becoming a typhoon, he is forced to turn off course and risk a rough landing at Okino-Tori-Shima, which western mariners know as Parece Vela. For the following few days, the Clipper tosses at anchor in the lagoon, but the lead hi-jacker is grateful for the cover of the storm. One of his crewmen believes that he may have seen a periscope beyond the reef, but the report is ignored.

Z'. 08/08/38 Monday: P.V.-Nagoya

In the late afternoon, after the storm abates, the Clipper departs from the small atoll and flies north to Nagoya, to land by night at Ise Bay and to seek shelter in a secured hangar. Several years later, when there are far fewer western eyes in Japan, she will be taken to Yokosuka. Later, in the summer of 1945, still in Japanese colors, she will be located by American military intelligence personnel and revealed to a few senior officers of the U.S. Pacific Commands. But a lid of secrecy will cover the truth: the *Zero*'s all-American engine could prove politically embarrassing.

4. APPENDIX

1. THE JACKSON / DALEY RESEARCH: Hard-won Pros and Calculated Cons

A writer, I have found, never escapes his work. Never.

—Ronald Jackson, *China Clipper*.

In the late 1970's, two writers were working independently on books about Pan American Airways. Ronald Jackson had intended to concentrate on the *China Clipper* and Ed Musick, but, after several years of research, found himself weaving *Hawaii Clipper*'s last flight into a specific history of PAA's Pacific adventure, focusing on the Japanese attempts to sabotage the revolutionary new American air route. Robert Daley, on the other hand, was creating a comprehensive history of PAA, focusing on its president, Juan Trippe, with little concern for Japanese interference outside of China. Both works, Daley's *An American Saga: Juan Trippe and his Pan Am Empire*, and Jackson's *China Clipper*, were published in 1980.

The *Hawaii Clipper* hi-jacking apparently took hold of Jackson's imagination, as he makes no bones about living "with images of people I never knew, from a past I did not live," and, considering that Jackson "thought about the people aboard the Hawaii Clipper constantly," one might well understand why, "more than once the weight of all of this overwhelmed" him. To those who have not been caught up in this old piracy and murder mystery, Jackson's apparent obsession may seem aberrant, but to one who has, indeed, been caught up in it, he was a kindred spirit.

In his efforts to provide solid contemporary evidence of the Clipper hi-jacking, however circumstantial it might be, and having no access to Juan Trippe's recollections (as told to Daley), or General Sutherland's report to Edna Wyman (as told to Hope Streeter), or Robert Nauroon's testimony (as told to Joe Gervais), nor a background in navigational mathematics, required to analyze the last flight, Jackson relied upon guidance from a few sources whom he trusted, and on public and private documents which outlined the Japanese efforts to sabotage the Clippers.

"One aspect of this story—the incidents of sabotage—eluded me for most of the four years and even now they are not complete...because many simply refused to talk or cautiously denied that they occurred. First the sabotage incidents are based on the oral confirmation by two men and one woman who said that they personally knew or had heard about these incidents. They are also based on FBI reports which took three years to obtain by the Freedom of Information Act, on letters from the National Archives, on news clippings and from reports found in Pan American Archives."

In setting the scene for a focus on sabotage, Jackson cites a front-page story in the *New York Times* of March 14, 1935. Leading with, "JAPAN IS OPPOSED TO PACIFIC LINE," the article quotes a Japanese Navy spokesman, who objected to Pan American's Pacific line because "the distinction between commercial airports and those for naval and military use was not clear. The landing places sanctioned could be converted into naval air bases whenever necessary..." In April of 1935, after Captain Ed Musick had completed a round-trip of the Hawaii leg of PAA's Pacific survey, Japan's *Harbin Nichi Nichi* stated, as Jackson notes, that "even if the route is restricted to commercial flights, who can assure that it would not be used for military purposes in case of emergency?...it looks like the line has a military purpose." Jackson goes on to quote Admiral Ernest J. King, then chief of the U.S. Navy Bureau of

Aeronautics, who, only a few weeks later, on May 15, stated that, "the requirements of Pan American Airways for a seaplane operating area are practically identical with the Navy requirements for an auxiliary operating area. These two developments should go hand in hand..." The Japanese had good reason to be concerned about PAA's developments in the Pacific, to be sure.

Jackson also notes that on May 22, 1935, the U.S. Navy Office of Naval Intelligence had informed Juan T. Trippe that, on May 17, a group of Japanese naval officers and civilian engineers had been granted permission to visit Sikorsky's aircraft plant at Bridgeport, Connecticut—but with the strict proviso that they not be permitted to examine the *S-42* flying boat. Rightly concerned that Japanese designers might incorporate features of the *S-42* into the new Kawanishi *H6K*, the Navy warned Trippe that these Japanese visitors, having been rebuffed at Sikorsky's plant, with respect to the *S-42*, might, during their PAA visit, attempt to examine one of PAA's *S-42*'s. The Navy "desired" that he refuse such a request. As Jackson points out, however, the Japanese managed to create, in the *H6K*, an improved version of the *S-42*.

On November 22, 1935, as PAA prepared to launch the maiden "Trip" of their trans-Pacific line, events occurred as if scripted for a Hollywood "B" movie. Jackson notes, drawing on FBI records, that, "two Japanese nationals secretly slipped aboard the China Clipper..." and were, "trying to miscalibrate the direction finder," when FBI agents apprehended both of them. Jackson notes that, "on the third day of the China Clipper's maiden flight, a CNAC [China National Aviation Corporation] transport, en route from Hong Kong to Tientsin, was intercepted by two Japanese fighters and forced down." It was not *shot* down, as the *DC-2*, the *Kweilin*, was later attacked, but, considering that PAA owned most of CNAC, this was surely an assault on PAA.

And on January 5, 1936, the *China Clipper*'s hull was slashed open by a submerged object, just prior to take-off from Alameda. According to Jackson, the Alameda base manager, Karl Leuder, found that, "on the bottom of the bay, several concrete pedestals with iron rods embedded in the base had been placed so that they stuck up just a few feet below the surface."

In discussing the *Samoan Clipper* disaster of January 11, 1938, Jackson fully presents the official version, but adds a footnote, never revealed before the publication of his *China Clipper*: "Late in the afternoon of January 29, Robert Thach, Pan Am's vice president and trouble shooter in Washington, met privately with E. A. Tarn, special FBI agent. A preliminary investigation conducted by Pan Am into the loss of the *Samoan Clipper*, Thach said, had revealed that the flying boat was destroyed not by a gasoline explosion but by a bomb which had been planted in the boat's hull." Ten days after this tragedy, PAA had again experienced acts of sabotage against the *China Clipper*, and Jackson thus quotes Tarn's note that, "Mr. Thach stated that the few people who know about these incidents are *virtually frantic* over the fact that sabotage is again being attempted upon their ships [italics added by the author]."

Most important, however, was Jackson's report of advance warning, which preceded the *Hawaii Clipper* hi-jacking: "In late June through early July, Acting Secretary of Commerce, Colonel J. Monroe Johnson had heard rumors that one of the Clippers might soon be sabotaged. The talk had been vague and the rumors died down, but the fear of a potential disaster lingered with Johnson through late July. Secretary Johnson was among the very few who had heard these rumors; certainly those flying aboard the Clippers knew nothing about them."

In fact, PAA flight crews were well aware of the potential for sabotage or attack by the

Japanese, but it is most curious that the *Hawaii Clipper* hi-jacking was presaged by a general warning, as was also the case with the Pan Am 103 bombing, over Lockerbie, Scotland, in 1988.

According to Jackson, a month after *Hawaii Clipper* was lost, a nervous former Marine, who had served at Sumay and done work at PAA's facility, was strangely able to apprise the FBI of Japanese espionage activities on Guam. The impression that one receives from Jackson's story is that the former Marine, himself, may have been somewhat closer to Japanese espionage activities, on Guam, than he cared to admit, in any detail.

Robert Daley, in his book, *An American Saga: Juan Trippe and his Pan Am Empire*, provided the details of Ed Musick's initial experience in jettisoning gasoline from an *S-42* (during the first, aborted, New Zealand survey, March 17, 1937), as well as the details of Leo Terletsky's first experience in dealing with an armed assault on a Clipper (Cuba, August, 1933). He also notes, with undisguised contempt, a body of "rumors" regarding the *Hawaii Clipper* hi-jacking:

"Almost immediately rumors were heard. Supposedly two Japanese gunmen had hidden in the tail, which explained why the plane had seemed to take off tail-heavy from Guam, remaining stuck to the water about twenty seconds longer than normal. In flight, after remaining hidden for some hours, the Japanese had come forward and hijacked the plane (the word "hijack" was not used in 1938 with regard to aircraft, and indeed the concept was unknown). The plane was then flown to Yokohama, where all aboard were killed. There was no lack of motives for such a hijacking: a Chinese passenger aboard was carrying a suitcase full of money for Chiang Kai-shek; the Japanese government had wanted the Pratt and Whitney engines to copy for their warplanes; Japanese nationals, who may or may not have been under official orders, had acted out of hatred of everything American."

Although Daley gave these "rumors" short shrift, they were far more than scuttlebutt, as is evident in his source notes, and in the sources: "The rumors: Trippe repeated them to the author at least three times, and at one time or another noted all of the details presented in the text. The dates of the interviews in which he returned to this subject were June 21 and 27, 1978, and May 16, 1979. He said that he had the story originally from Admiral J. T. Tower who joined Pan Am as a vice president after World War II. Every Japanese *Zero* examined by U.S. technicians during and after the war bore the same serial number on their magnetos, the same serial number as the magneto on one of the engines of the Hawaii Clipper, said Tower to Trippe, according to Trippe. On Aug. 26, 1970, he told the story to a man named Harvey Katz, who made a memo of the conversation; this memo is to be found in PAF 10.10.00; and he also told it to Langewiesche in 1957." But Daley had left out a few critical details:

One of the U.S. Navy's first aviators, John Henry Towers won his Navy wings in 1911, served as assistant director of naval aviation in WWI and commanded the Navy's famous transatlantic flight in 1919. He commanded the aircraft carriers *Langley* and *Saratoga* between the wars, and then served, during the Pacific War, as the Deputy Commander in Chief, Pacific Area, succeeding Nimitz as the Commander in Chief, Pacific Fleet and Pacific Ocean Areas, for a short time, in November of 1945. Whatever Admiral John Henry Towers may have told Juan Trippe, then, can hardly be dismissed as "rumor."

It has occurred to the author that Robert Daley's 1980 book may have served, in part, as a deliberate counter to the Jackson book, whose long research phase would have made such a disinformation assault quite feasible.

2. THE GERVAIS DOCUMENT "PACKAGE:" Preservation through Dissemination

In November of 1964, Major Joseph L. Gervais, USAF (Ret.) took his quest for Amelia Earhart to Truk to check out a twin-engine aircraft, wrecked in a mangrove swamp. Although Joe's search for Earhart would later gain fame (and infamy) in Joe Klass' controversial book, *Amelia Earhart Lives*, his 1964 visit to Truk Atoll (now Chuuk State) yielded little more than a few photographs of an old beached seaplane and a chilling tale of the primitive interment of fifteen Americans. As Joe related the details to me in 1987, he had employed, as a guide, one Robert Nauroon, a Trukese of Franco-Micronesian descent, who, after the swamp wreck was identified as a WWII Japanese "*Betty*" bomber, showed Joe a concrete slab on Dublon Island, where, in the summer of 1938, as he told Joe, he and Taro Mori, a Trukese of Nippo-Micronesian descent, had entombed fifteen men, said to be Americans:

Under contract to the Imperial Japanese Navy, Nauroon and Mori had brought their crews to Dublon Island, early one morning in the late summer of 1938, to pour a concrete foundation slab for a planned naval hospital, to serve the Fourth Fleet. Nauroon had been merely surprised to find a Japanese military presence at the site, but was soon shocked, when he saw that the foundation forms enclosed the bodies of fifteen men.

According to Joe's account of Nauroon's story, the bodies were arranged face down, in three rows of four each and one row of three, in the northern half of the shallow excavation, which was about thirty feet wide by sixty feet long, overall. Nauroon recalled that many of the bodies wore dark uniforms and that he had been made aware that they were Americans. The job promised to be most unpleasant, but it was dangerous to oppose the Japanese, so the crews worked swiftly, and by noon, the bodies had been covered and the slab had been finished.

In his apparent belief that Joe Gervais might somehow accomplish the recovery of the fifteen remains, Nauroon repeatedly assured Joe that all fifteen men had been dead when immured in the concrete, but added that, when recovered, there would be found "no marks of death upon them," a phrase which Joe took to mean that the fifteen had not been shot or beheaded, but had been killed by other, less apparent means. However, in a subsequent conversation with Taro Mori, who seems to have been more ill at ease than Nauroon had been (as Joe indicated), Mori confirmed all that Nauroon had told him, adding that reinforcing rod or wire had been required "to keep the bodies down."

Whether these details indicate that measures had been taken to keep the dead bodies from floating to the surface of the wet cement, or to keep living men from attempting to rise above a suffocating tide of concrete, will be unknown until the remains are recovered, but these details are of interest, in that they seem more likely to be a product of the two men's experience than of their imagination.

And, although he spoke less freely of his own activities than of what he had observed, Nauroon, according to Joe, had been specific about the year, the time of year, the number of bodies, their indicated nationality, the dark uniforms (consistent with PAA crews), the arrangement in the excavation and the location of the slab. He apparently identified none of the Japanese by name (there still being fear of the Japanese on Truk in 1964, as there had been in 1946, and before), and seemed intent only upon easing his own sense of responsibility for the burials. If this was his motivation, as it seems to have been, it is probable that he had already provided a report to the U.S. war crimes investigators, to the later U.S. Navy administrators, or to the still later civilian officials of the Interior Department and of the Trust Territory.

In 1980, following the publication of Ronald Jackson's *China Clipper*, Joe realized that the story he had heard on Truk, in 1964, dovetailed perfectly with the details of the *Hawaii Clipper* disappearance (with which he had not been familiar in 1964). And, on July 6, 1980, the *New York Times* and the *Los Angeles Times* carried an Associated Press story by Patrick Arnold, outlining Joe's 1964 experience and indicating Joe's interest in reopening the Clipper case.

However, Joe had also provided photos of a nearly disintegrated seaplane, which he had found on Moen Island, and further interest in any fresh investigation was swiftly quashed, due primarily to Joe's identification of the wreckage as that of the Clipper. For reasons of his own, Joe let the Clipper go, but not before he had assembled and distributed a document "package," which combined the August 2, 1938, PAA *Engineering Report* and the November 18, 1938, *Preliminary Report* of the combined Commerce and Air Safety Boards, with documents relating to his 1964 trip to Truk. One such "package" was provided to John F. Luttrell, of Atlanta, who provided a copy to the author in 1987. Subsequent conversations with Joe Gervais elicited the full details of Robert Nauroon's story, which are recounted in this book. The Luttrell copy is herein reproduced, in part, with the kind permission of Joe Gervais, and will be seen to bear John F. Luttrell's initials (JFL), as well as his comments and notes.

It will be noted that the package contains no detailed documentation of Robert Nauroon's story, but it is a story that does not have to be heard twice to be remembered. And, as unpleasant as it is to discuss the details, it is far more difficult to commit them to paper.

ORIGINAL "PACKAGE" CONTENTS

```
A. The China Clipper       (PAA Photo)
B. PAA Engineering Report     (5 pp.)
C. PAA Engineering Memo       (1 p.)
D. CAA Press Release          (2 pp.)
E. ASB Preliminary Report    (17 pp.)
F. Passenger and Crew List    (Notes)
G. Source Material         (Clippings)
H. Source Material             (Notes)
I. Source Material (Contacts, Truk)
J. Source Material (Contacts, U.S.)
K. Associated Press Story   (7/6/80)
L. Continental Hotel(Advertisement)
M. Visitor Permit     (Cover Letter)
N. Island Names          (Truk Atoll)
O. Visitor Permit 2946 (Trust Terr)
P. Truk Atoll       (WWII JICPOA Map)
Q. Dublon Island        (USCGS Chart)
R. Hospital Slab    (Photo and Text)
S. Personnel Barge  (Photo and Text)
T. Aircraft Wreck   (Photo and Text)
U. Aircraft Wreck   (Photo and Text)
V. Pan Am Business Card    (Contact)
W. Pan Am Crew Conference    (Photo)
```

NOTES

Although the entire document "package" has been reproduced, it will be noted that some published material has been excised, while documents critical to the analysis of the hijacking have been reproduced in full, as Joe packaged them. The ASB Report is included for its wealth of data, but the vital record is Priester's PAA *Engineering Report*.

PAA's August 2, 1938, *Engineering Report* was the work of the Dutch Chief Engineer, Andre Priester (possibly with the help of an assistant, better skilled in English). It is the sole source for the four positions reported by *Hawaii Clipper* during the two hours prior to her final position (which had been widely reported by the press). Although his reason for recording these five positions, alone, is unknown, the five had been logged only by Panay, and it may be that, having perceived their importance, he intended that they be prominently preserved in PAA's archives. Note the Pan American Foundation (PAF) file numbers handwritten on the first page.

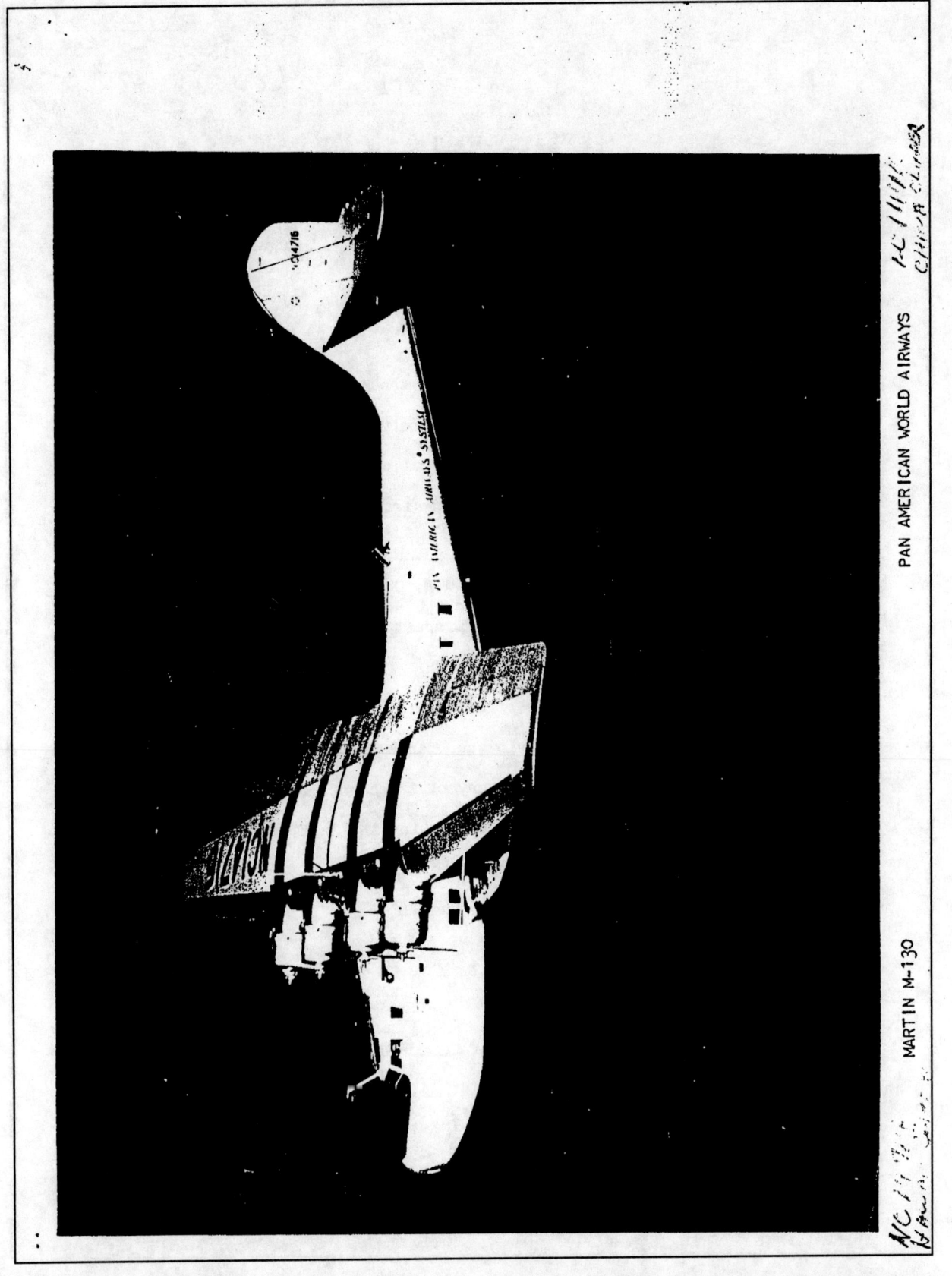

A. The *China Clipper* (PAA photo)

This photograph, although not of *Hawaii Clipper*, was included to provide a graphic, recognizable representation of the Martin *M-130* flying boat.

[Handwritten at top: PAN AM FILE NOS.? JFL   X10 10.00 (illegible) 30.20.02 50.20.0P JFL]

HAWAIIAN CLIPPER

Engineering Department
August 2, 1938

1. **Aircraft Log**

    (a) The scheduled flight of the Hawaiian Clipper westbound from Alameda to Guam was entirely normal in every respect and without incident. No unusual or severe conditions of weather or sea were encountered and aircraft and engines functioned normally.

    (b) Departure Guam 5:39 A.M. Guam Time. Good sea conditions at Guam for take-off. Aircraft fueled for 2600 mile range - route mileage 1589 to Manila.

    (c) From the time of departure, position reports were received every thirty minutes in accordance with standard procedure. Nothing unusual was indicated in any of these reports. There follow the position reports received during the last two hours. All positions since departure from Guam were computed by dead reckoning with the exception of one sun line.

    0200 GCT (9 pm est - 12 n Guam)

    Latitude 12.20
    Longitude 134.26
    Weather - squalls all around

    0230   Latitude 12.00
            Longitude 133.36

    0300   Latitude 12.05
            Longitude 133.28
            Clouds 10/10 above 10/10 below
            Average altitude 10,000

[Handwritten annotations in margin: "65 NAUTICAL MILES JFL IN 1 HR STRANGE" and "9 N-MILES or 30 MIN. JFL"]

B. PAA *Engineering Report* (p. 01/05)   *Engineering Report* of August 2, 1938, from PAA Chief Engineer, Andre Priester; the sole source for four of the Clipper's five final position reports.

-2-

[handwritten: JFL 0300 LAT. 12.05 LONG. 133.28]

[handwritten: 160-170 N.M. DURING THIS 1 HOUR]

    0330 GCT   Latitude 12.15
                     Longitude 131:37

    0400   Latitude 12.27
                     Longitude 130.40
                     Clouds 10/10 stratocumulus (above)
                             10/10 cumulus (below)

                     Altitude 92.00
                     Rain

        (d) Details of position report 0400 GCT.

The 0400 GCT position report was filed at 0403 by the Captain and transmission was completed at 0411. As soon as the transmission of the position report was completed, the ground station radio operator advised the aircraft that he had a weather sequence to transmit. The aircraft advised that he was experiencing rain static, to wait. The ground operator waited one minute and then called the aircraft again, but was unable to raise the operator. An attempt was made to use the Direction Finder, but no signal could be heard from the aircraft.

2. **Weather**

        The weather forecast prior to departure from Guam indicated good weather for the entire route, some clouds, scattered showers of light intensity and normal west winds varying from 12 to 14 miles per hour. The aircraft reported scattered squalls and rain static was experienced in connection with radio transmission. At the time of the last position report the aircraft was flying at an altitude of 9200 feet with a solid

B. PAA *Engineering Report* (p. 02/05)

Notes, and the initials, "JFL," are those of John F. Luttrell, of Atlanta, who provided the copy of Joe Gervais's document package, early in 1987.

-3-

layer of stratocumulus clouds above and solid cumulus below, air moderately rough, and experiencing rain. The U. S. Meigs, which reached this last position several hours later, reported smooth sea, light winds and good visibility.

During the previous two hours of flight the Captain had changed his course frequently, indicating that he was flying around squalls. There has been no evidence that there existed any severe meteorological phenomena in the air at the time.

3. <u>Rescue Measures</u>

Attempts were made to contact the aircraft between 0400 and 0430 and when it did not report the next required radio contact at 0430 GCT, the steamer U.S. Meigs enroute to Guam was immediately called. The U. S. Meigs, which was approximately 90 miles from the last position reported by the aircraft, immediately proceeded to the site of the last position and stood by to begin a search at daylight. The last position of the aircraft was approximately 565 miles from Manila on the regular southern course between Guam and Manila. This position was also approximately 300 miles east of the nearest island of the Philippines. The following were also dispatched immediately to aid in the search and the areas which they covered are indicated on the chart.

    5 Long Range Army Bombers

   13 Navy Surface Vessels, including three Destroyers and an Aircraft Tender

    2 Navy Amphibions.

B. PAA *Engineering Report* (p. 03/05)

-4-

The search continues to date and the only evidence obtained was the report of an oil slick by the U. S. Meigs which was reported to contain gasoline, as well as oil. This was located about 50 miles from the last reported position of the aircraft.

4. Crew and Passengers

Crew:
Captain Leo Terletzky
First Officer ... M. A. Walker
Second Officer... G. H. Davis
Third Officer ....J. N. Sauceda
Fourth Officer....J. W. Jewett
Engineer Officer..M. L. Cox
Asst. Engineer Officer... T. B. Tatum
Radio Officer.... W. McCarty
Flight Steward... I. Parker

Passengers:
E. E. Wyman.............. Bronxville, N.Y.
Fred Meier............... Washington, D. C.
Earl McKinley............ Washington, D. C.
Howard French............ Portland, Oregon
K. A. Kennedy............ Palo Alto, Cal.
Choy Wah-Sun............. Jersey City, N.J.

5. Conclusion

The entire flight prior to the last reported position was carried out strictly in accordance with Pan American Airways established long range flight procedure. No difficulties, or unusual incidents, had been reported with respect to the aircraft, engines, radio, or the crew. The aircraft and engines are the most modern in long range operation today and have been successfully operated in Trans-Pacific service for over three years without serious incident. The crew was thoroughly trained and had had extensive Trans-Pacific experience. They were carrying out their duties on this flight in accordance with established procedure in every respect. Captain Terletzky has had over ten years flight experience in the Eastern and Pacific Divisions

B. PAA *Engineering Report* (p. 04/05)

-5-

of Pan American Airways and has had almost two years experience in Trans-Pacific service. First Officer Walker was a Junior Pilot thoroughly qualified in instrument flying, navigation, engineering, meteorology, and long range communication. The other members of the crew were thoroughly trained and had gained extensive experience on previous assignments to Trans-Pacific flight crews. There is no direct evidence as to what occurred after the last position report. Circumstantial evidence indicates that the aircraft was violently thrown out of control during the minute which intervened between the last position report and the ground operator's unanswered call. The extensive experience of the crew, and the excellent condition of the aircraft and engines makes it difficult to concede that there was any failure of the aircraft or personnel.

A. A. Priester.

B. PAA *Engineering Report* (p. 05/05)

cc: Vice President C. L. Rihl Vice President J. C. Cooper, Jr.
    Vice President E. E. Young Vice President H. M. Bixby
                                              Assistant Comptroller
  Division Managers: Chief Communications Engineer
  Operations Managers: Resident Engineer   November 22, 1938
  Division Engineers:
    Eastern Division
    Western Division   Chief
    Panair do Brazil
    Pacific Division   Engineering
    Atlantic Division
    Northwest Division   New York
  Maintenance Engineer- Rio

C.A.A. Report - Hawaiian Clipper

                                  There is attached, for your

information, a copy of the Civil Aeronautics Authority

preliminary report of November 18, 1938 of the

investigation of the disappearance of the Hawaiian

Clipper on July 29, 1938.

                                                   A. A. Priester.

Enclosure.

C. PAA Engineering Memo (p. 01/01)   Cover letter for internal PAA distribution of the November 18, 1938 CAA Press Release and the ASB *Preliminary Report* on the Clipper loss.

FOR IMMEDIATE RELEASE   November 18, 1938.

CIVIL AERONAUTICS AUTHORITY
WASHINGTON, D. C.

The Air Safety Board made a preliminary report to the Civil Aeronautics Authority today on the disappearance of the Hawaii Clipper with six passengers and a crew of nine aboard last July 29 between Guam and Manila on a regularly scheduled flight across the Pacific.

The document reveals that no authentic trace of the airplane ever has been found and that, as a result, those who investigated its disappearance are unable to ascribe any probable cause for the disaster that is presumed to have overtaken the big flying boat. Pending the discovery of some concrete evidence as to the fate of the Hawaii Clipper, the investigation remains in an open status.

In transmitting the report, the Safety Board which consists of Colonel Sumpter Smith, Chairman, and Thomas O. Hardin, sets forth that it was prepared and submitted by a Board of Investigation originally constituted by the Secretary of Commerce, prior to the creation of the Civil Aeronautics Authority and the Air Safety Board. The Secretary of Commerce directed that this board be relieved of its obligation to report to him, continuing its investigation and turning over the results thereof to the Air Safety Board of the Civil Aeronautics Authority. Members of the board were R. D. Hoyt, Aeronautic Inspector, Phil C. Salzman, Airline Maintenance Inspector, and W. T. Miller, Airway Superintendent, International Section.

The report discloses that an oil slick discovered by the United States Army Transport Meigs near the last reported position of the Hawaii Clipper the day after its disappearance was proved by chemical analyses and other tests to have no connection whatever with the airplane
18425

D. CAA Press Release (p. 01/02)   November 18, 1938; from the Civil Aeronautics Authority, announcing the *Preliminary Report* of the Air Safety Board, regarding the Clipper loss.

-2-

though at the time of its discovery this was generally regarded as a likely clue to the flying boat's fate. It refuses to dismiss, however, as without possible bearing on the disappearance of the airplane, a report from the Island of Lahuy that a large airplane was heard flying above the clouds about 3 P.M., Manila Time, on July 29. Lahuy is a small island lying east of and close to the southeastern tip of Luzon, and is directly on the course from the Clipper's last reported position to Manila. The report points out that at normal cruising speed the flying boat should have reached that vicinity at approximately 3:46 P.M., Manila Time. Luzon is the largest island of the Philippine group and while its southern portion, and the other large islands in the vicinity, are moderately well settled, they have large areas of tropical jungle and mountainous territory rising to an altitude exceeding 7000 feet.

A careful aerial search was made of Lahuy and nearby areas after it had been ascertained that no Army, Navy, or private aircraft were in that vicinity the day the clipper disappeared, but with negative results. In its summation, the report says, however, that "the Board is not prepared to say that the aerial search of Lahuy and other areas in the Island can be considered as conclusive".

Attached is a copy of the full report.

AIR SAFETY BOARD
of the
CIVIL AERONAUTICS AUTHORITY

****

PRELIMINARY REPORT OF INVESTIGATION OF
THE DISAPPEARANCE OF AN AIRCRAFT OF
PAN AMERICAN AIRWAYS, INCORPORATED,
IN THE VICINITY OF LATITUDE 12°27'
NORTH, LONGITUDE 130° 40' EAST,
ON JULY 29, 1938.

In an order dated August 17, 1938, issued by the Secretary of Commerce, pursuant to the provisions of the Air Commerce Act of 1926, as amended, R. D. Hoyt, Aeronautical Inspector, Phil C. Salzman, Airline Maintenance Inspector, and W. T. Miller, Airways Superintendent, International Section, were designated as a Board to investigate the facts, conditions and circumstances surrounding the disappearance of the Hawaii Clipper, an aircraft of Pan American Airways, Incorporated, while flying westbound between Guam and Manila, P. I., on July 28, 1938, and to make a report thereon. Colonel Sumpter Smith and Mr. Thomas O. Hardin of Washington, D. C., were designated as advisory members of the Board.

The Board convened at Alameda, California on August 18, 1938, and promptly instituted an investigation. After a preliminary survey, the Board deemed it inadvisable to hold a public hearing at that time and a report to that effect was made to the Secretary of Commerce.

Thereupon the Secretary of Commerce directed that the Board be relieved of that duty, and directed that it continue the investigation and deliver its report of such investigation to the Air Safety Board of the Civil Aeronautics Authority, as part of the records of the Department of

E. ASB *Preliminary Report* (p. 01/17)   Report of the original Department of Commerce Board, released by the Air Safety Board under the aegis of the new Civil Aeronautics Authority.

- 2 -

Commerce transferred to the Civil Aeronautics Authority pursuant to Executive Order 7959 dated August 22, 1938.

The Board, therefore, proceeded with the investigation and reports as follows:

On July 29, 1938, at 0411, Greenwich Civil Time, the Captain of the Hawaii Clipper, an aircraft of Pan American Airways, Incorporated, flying westbound between Guam and Manila, P. I., in regular scheduled foreign air commerce, sent a routine radio position report giving their 0400 G.C.T position as being Latitude 12°27' North, Longitude 130°40' East. This position is approximately 582 nautical miles east southeast of Manila, P. I. The position report was received by the radio operator at Panay, P. I., who was at that time standing radio guard over the operation. Panay acknowledged this report and requested the Clipper to stand by for a routine weather sequence report. The radio operator on the Clipper requested that the report be held up for a minute on account of rain static. Following that request, the radio operator at Panay was unable to obtain acknowledgement of messages sent the Clipper, and since that date no reasonable explanation for its disappearance has been found.

The date and time mentioned above correspond to 12:11 P.M. of July 29th, Local Time at Manila, P. I., and 11:11 P.M. of July 28th, Eastern Standard Time.

At the time of the aircraft's disappearance, it was being flown in scheduled foreign air commerce under a letter of Authority dated September 19, 1936, and carried passengers, mail and express. The crew consisted of the Captain, First Officer, Second Officer, Third Officer, Fourth Officer, Engineer Officer, Assistant Engineer Officer, Radio Officer and Flight Steward.

The Captain, Leo Terletzky, held a current and appropriate Federal Pilot's Certificate of Competency, and the required ratings for the flight involved. His last physical examination as required by the Department of Commerce was taken on April 20, 1938, with a subsequent physical examination taken on July 21, 1938, in conformity with Company regulations for all flight personnel, both of which showed him to be in satisfactory physical condition. Captain Terletzky's record indicates that he had a total of 9,200 hours flying experience, of which 1626 hours had been in trans-Pacific operation. His total flying experience in the Martin, Model 130 Flying Boat, was 1614 hours. His total flying time in the preceding year was 826 hours, and his flying time in the month preceding the start of the subject trip was 6 hours.

First Officer Mark A. Walker held a current and appropriate Federal Pilot's Certificate of Competency, and the required ratings for the flight involved. His last physical examination as required by the Department of Commerce was taken on February 28, 1938, with a subsequent physical examination taken on July 21, 1938, in conformity with Company regulations for all flight personnel, both of which showed him to be in satisfactory physical condition. His record showed that he had a total of over 1900 hours flying experience, 1575 hours of which had been in trans-Pacific operation.

Second Officer George M. Davis held a current and appropriate Federal Pilot's Certificate of Competency (License), and the required ratings for the flight involved. His last physical examination as required by the Department of Commerce was taken on February 14, 1938, with a subsequent physical examination taken on July 21, 1938, in conformity with Company regulations for all flight personnel, both of which showed him to be in satisfactory physical condition. Pilot Davis' record showed that he had a total of 1650 hours flying experience, 1080 hours of which had been in trans-Pacific operation.

E. ASB *Preliminary Report* (p. 03/17)

Third Officer Jose M. Sauceda held a current and appropriate Federal Pilot's Certificate of Competency (License), and the required ratings for the flight involved. His last physical examination as required by the Department of Commerce was taken on January 7, 1938, with a subsequent physical examination taken on July 21, 1938, in conformity with Company regulations for all flight personnel, both of which showed him to be in satisfactory physical condition. His record showed that he had a total of 1900 hours flying experience, 570 hours of which had been in trans-Pacific operation.

Fourth Officer John W. Jewett held a current and appropriate Federal Pilot's Certificate of Competency, and the required ratings for the flight involved. His last physical examination as required by the Department of Commerce was taken on June 23, 1938, with a subsequent physical examination taken on July 21, 1938, in conformity with Company regulations for all flight personnel, both of which showed him to be in satisfactory physical condition. His record showed that he had a total of 2000 hours flying experience, 428 hours of which had been in trans-Pacific operation.

Engineer Officer Howard L. Cox held a current and appropriate Federal Aircraft and Engine Mechanic's Certificate of Competency. His last physical examination, taken on July 21, 1938, in conformity with Company regulations for all flight personnel, showed him to be in satisfactory physical condition. His record showed that he had over 944 hours flying experience in trans-Pacific operation.

Assistant Engineer Officer T. B. Tatum held a current and appropriate Federal Aircraft Mechanic's Certificate of Competency. He had boarded the aircraft at Honolulu and was on a training flight.

Radio Officer William McCarty held a Federal Radio License, First Class. His last physical examination, taken on July 21, 1938, in conformity with Company regulations for all flight personnel, showed hime to be in satisfactory physical condition. His record showed that he had 1352 hours radio experience in trans-Pacific operation.

The Flight Steward was Ivan Parker. His last physical examination, taken on July 21, 1938, in conformity with Company regulations for all flight personnel, showed him to be in satisfactory physical condition. His record showed that he had 1200 hours flying experience in trans-Pacific operation.

The passengers were:

Edward E. Wyman, 15 Edgewood Lane, Bronxville, New York.

K. A. Kennedy, 2534 Mountain Boulevard, Oakland, California.

Dr. Earl B. McKinley, 2101 Connecticut Ave., Washington, D. C.

Fred C. Meier, U.S. Department of Agriculture, Washington, D. C.

Major Howard French, 2560 North East 30th Ave., Portland, Oregon.

Choy Wah Sun, 2928 Hudson Boulevard, Jersey City, N. J.

The aircraft, a Martin, Model 130 Flying Boat, was owned and operated by the Pan American Airways, Incorporated, whose principal place of business is 135 East 42nd Street, New York City, N. Y. It was delivered to the operator by the manufacturer at Baltimore, Maryland, on December 24, 1935, and was last certificated as air-worthy on July 23, 1938. It bore Federal Certificate of Airworthiness (License) Number NC-14714.

On September 19, 1936, the Department of Commerce granted authority to the operator to utilize this aircraft for the carriage of passengers and cargo in scheduled service between Alameda, California, and Manila, P. I., with intermediate stops at Honolulu, Midway, Wake, and Guam, at a maximum gross weight of 52,000 pounds. The specified conditions included a mandatory provision for the jettisoning (dumping) of sufficient fuel to reduce the gross weight to 42,000 pounds for landing. While in trans-Pacific service, the

E. ASB *Preliminary Report* (p. 05/17)

aircraft completed 35 round trips between Alameda, California, and Manila, P. I.

The total flying time of the Hawaii Clipper prior to departure from Alameda on Trip No. 229 was 4751:55 hours. It had flown an additional 55:58 hours when last reported between Guam and Manila.

The engines, which are overhauled after each three round trips, or at approximately 390 hours, had the following hours of service, since last overhauled, at the time of departure from Alameda:

1. Left outboard engine   3:48 hours
2. Left inboard engine   137:26 hours
3. Right inboard engine   3:48 hours
4. Right outboard engine   273:25 hours

The propellers, which are overhauled at the same periods as the engines, had the following hours of service, after overhaul, at the beginning of Trip No. 229:

1. Left outboard propeller   140:41 hours
2. Left inboard propeller   140:41 hours
3. Right inboard propeller   3:48 hours
4. Right outboard propeller   140:41 hours

The operation and maintenance history of this aircraft has been normal in all respects in comparison with other flying boats operated by the Pan American Airways Company between San Francisco and Manila.

All changes recommended by the Glenn L. Martin Company had been completed on the Hawaii Clipper and there were no incomplete overhaul items carried over to the next arrival at Alameda.

The trip in question was known as Trip No. 229, scheduled from Alameda, California, to Manila, P. I. Departure from Alameda was made on July 23,

1938, the flight arriving at Honolulu on July 24th. On July 25th the trip was continued from Honolulu to Midway on schedule, arriving there July 26th. A scheduled departure was made from Midway on July 26th and the flight arrived at Wake on July 27th. It departed from Wake on July 27th and arrived at Guam at 0555 G.C.T. on July 28th. Departure from Guam was made at 1939 G.C.T. (3:39 A.M. Manila Time) on July 28, 1938, the aircraft actually leaving the water 29 minutes later. Sunset at Manila, the destination, was at 6:29 P.M., Manila Time, thus giving 14 hours and 50 minutes of daylight for the flight. The flight time analysis indicated that in standard cruising at a density altitude of 7800 feet, the flight would take 12 hours 30 minutes. There were 2550 gallons of gasoline on board, and 120 gallons of oil. This amount of gasoline was calculated to provide a cruising endurance of 17 hours and 30 minutes at cruising R. P. M. under the conditions stated above. The aircraft carried six passengers, 1138 pounds of cargo, and was loaded to a gross weight of 49,894 pounds at the time of takeoff.

At 0330 G.C.T., 41 minutes before the last radio contact, there remained aboard the aircraft 1420 gallons of gasoline which is sufficient for 10.1 hours of normal cruising. This would indicate that there was ample fuel aboard to reach Manila and continue the flight for several hours thereafter.

The records show that the trip from Alameda to Guam was uneventful. Weather conditions were better than average, and each leg of the trip was accomplished on schedule. Between Midway and Wake, the course followed was considerably to the South of the regular route due to bad weather along the route and to the north. This procedure is normal, as regular alternate routes are laid out for each leg of the Pacific operation. The route selected for any trip is based on weather forecasts made prior to departure, and may be changed enroute at the discretion of the Captain.

E. ASB *Preliminary Report* (p. 07/17)

The records also show that the ordinary routine radio contacts were maintained between the aircraft and the shore stations on guard on each leg of the trip. Company procedure requires that a constant guard be maintained by radio stations at the point of departure and destination, at all times while the aircraft is in flight.

Prior to departure from Alameda, the aircraft was inspected, and routine service procedure, known as "long airplane service" and "long engine service", was carried out. The "shore run up report" and engineering flight tests showed satisfactory operation. The Company's routine procedure calls for a test flight of at least three hours on the day preceding departure from Alameda. This flight is conducted by the same crew scheduled to make the trip.

During this flight an emergency landing is made and a routine "abandon ship" drill is carried out. This drill consists of inflating the life raft and getting it over-side, with the crew, emergency equipment and rations aboard. An emergency radio is set up and communication with shore stations is established. All members of the crew have a definite assignment during this maneuver.

During over-night stops at Honolulu, Midway, Wake and Guam the aircraft was serviced and inspected in accordance with routine procedure which calls for "over-night airplane service" and "over-night engine service". The former requires inspection of vital parts and the addition of fuel, oil, water and auxiliary fluids. The latter requires a detailed inspection of the engines, and their controls and installations. No significant irregularities were reported by the Flight Engineer or detected by the respective airport Chief Mechanics.

-9-

At Wake Island, the Hawaii Clipper met the Philippine Clipper which was eastbound, both aircraft making a scheduled overnight stop at that point. Testimony from members of the crew of the Philippine Clipper indicates that the entire crew of the Hawaii Clipper were in the best of spirits and that they reported a comfortable and normal trip up to that time.

The trip weather forecast, prepared at 1322 G.C.T., July 28, on the basis of which the aircraft was dispatched from Guam, was as follows:

	Zone 1 Long 145° to 140° East	Zone 2 Long 140° to 130° East	Zone 3 Long 130° to 120° East
Weather Conditions	Scattered showers	Scattered showers	Scattered Thunder showers
Visibility	Good	Good	Good
Ceiling	2,500 ft.	2,500 ft.	3,000 ft.
Cloud tops	8,000 ft.	8,000 ft.	8,000 to 10,000 ft.
Sea Conditions	Slight	Slight	Slight
Base of Upper Clouds	4 CS AS 14,000 ft.	8 CS AS 14,000 ft.	6 AS 14,000 ft.
Wind at 4,000 ft.	SW 14	WSW 10	W 12
Wind at 7,800 ft.	WSW 14	WSW 10	W 12
Wind at 11,500 ft.	WSW 12	W 10	WNW 10

Remarks: Thunder clouds all zones 14,000 to 16,000 feet.
          Widely scattered thunder showers over Archipelago.

The foregoing forecast was issued by the Company Meteorologist, whose headquarters is in Manila, P. I.

Local weather at Guam at 1939 G.C.T., the time of departure from that point, was as follows:

Weather Conditions	-Partly Cloudy
Visibility	-Unlimited
Amount of Clouds	-2/10 - lower 5/10 total
Height of lower clouds	-2,000 feet
Weather for past hour	-Partly Cloudy
Wind	-WSW 6
Barometer	-29.83
Temperature	-80° F
Water conditions	-Moderate

E. ASB *Preliminary Report* (p. 09/17)

- 3 -

The local weather at Cavite, P. I., at this time, was as follows:

      Weather Conditions   -Partly Cloudy
      Visibility   -Unlimited
      Amount of Clouds   -1/10
      Height of lower clouds   -2,500 feet
      Weather for past hour   -Partly Cloudy
      Wind   -SW 4
      Barometer   -29.76
      Temperature   -81° F
      Water Conditions   -Smooth

Guam upper air at 1830 G.C.T. was as follows:

    Surface   - WSW 6
    1,000 ft. - WSW 10
    2,000 ft. - WSW 14
    3,000 ft. - WSW 17
    4,000 ft. - WSW 18
    5,000 ft. - WSW 19
    6,000 ft. - WSW 21
    7,000 ft. - WSW 18
    8,000 ft. - WSW 19
   10,000 ft. - WSW 17

Cavite upper air at 1700 G.C.T. was as follows:

    Surface   -SW 3
    1,000 ft. - W 10
    2,000 ft. - W 16
    3,000 ft. - W 16
    4,000 ft. - W 15
    5,000 ft. - WSW 21
    6,000 ft. - W 11
    7,000 ft. - W 12
    8,000 ft. - WSW 20
    9,000 ft. - WSW 12

Radio facilities at Guam consist of a Company Station KMBG. Other stations standing guard on the operation between Guam and Manila were Panay KZDY and Manila KZBQ. In addition, other Pan American Airways radio stations in the Philippine Islands were standing watch. The aircraft was equipped with two independent transmitters, either of which could be used for communicating on assigned frequencies or on the international distress frequency of 500 kilocycles. Radio communication was maintained in a normal manner up to 0411 G.C.T. At that time a routine report was sent by the Clipper giving their position at 0400 G.C.T. (12 Noon Manila Time).

The message was as follows (all communications are in code and messages quoted are interpretations):

"Flying in rough air at 9100 feet. Temperature 13° centigrade. Wind 19 knots per hour from 247°. Position Latitude 12°27' N, Longitude 130°40' E dead reckoning. Ground speed made good 112 knots. Desired track 282°. Rain. During past hour cloud conditions have varied. 10/10ths of sky above covered by strato cumulus clouds, base 9200 feet. Clouds below, 10/10ths of sky covered by cumulus clouds whose tops were 9200 feet. 5/10ths of the hour on instruments. Last direction finder bearing from Manila 101° True."

The radio operator at Panay acknowledged this message and indicated that he wished to transmit weather sequence reports based on observations compiled at 0400 G.C.T. by the Philippine Stations and relayed to him in accordance with Company procedure. The radio operator on the Clipper replied as follows:

"Stand by for one minute before sending as I am having trouble with rain static."

The radio operator in Panay again called the Clipper at 0412 G.C.T., giving him the weather sequences. This message was not acknowledged. The radio operator at Panay continued calling the Clipper until 0415 G.C.T., at which time he sent the Clipper's 0400 G.C.T. position report to Manila. At 0435 G.C.T., the Communications Superintendent, Pacific Division, Alameda, Calif., was notified of the failure of communication between ground stations and the Clipper. All Philippine stations were requested to stand by on emergency communication frequencies at 0449 G.C.T. This is a standard emergency procedure which is followed whenever there is an interruption of communication between shore stations and the aircraft on which they are standing guard. Such interruptions occur from time to time and while they occasion no alarm, the emergency procedure is put into effect as a matter of routine.

At 0900 G.C.T. (5 P.M.) the Clipper was due at Manila. Up to that time, while some alarm had been felt, it was believed that she might be proceeding to her destination with the radio out of commission. As soon as she was overdue, the Naval Commandant, at the request of Pan American Airways, ordered all Navy vessels at Manila, to stand by for maneuvers. At 1030 G.C.T. (6:30 P.M.) all Navy vessels were ordered to refuel and prepare to put out to sea. At 1600 G.C.T. (midnight) 13 vessels were under way to begin a search for the missing aircraft.

The nearest surface vessel at the time was the United States Army Transport "Meigs", which at 0411 G.C.T. was approximately 103 miles west northwest of the Clipper's last reported position and headed east. Upon receipt of a report that shore stations had lost contact with the Clipper, the Meigs altered her course at 0740 G.C.T. (3:40 P.M.) Manila Time, and proceeded to the Clipper's last reported position, arriving in that vicinity at 1400 G.C.T., July 29 (10 P.M.) Manila Time.

A search was conducted in the vicinity for approximately three hours. Following that time, the search was conducted over various areas throughout that night, and the following day, in accordance with instructions from Pan American Airways. At 0910 G.C.T. (5:10 P.M.) Manila Time on July 30, an oil slick was discovered about 28 miles south southeast of the Clipper's last estimated position. This slick was variously estimated by officers of the Meigs to be from 500 to 1,500 feet in diameter and roughly circular in shape.

The area in the cicinity of the slick was carefully searched and a small boat was put over in charge of Second Officer J. A. Harrington, for the purpose of obtaining samples of oil from the slick. Due to the limited amount of time before darkness, only a small sample of oil was obtained. The Meigs remained hove to during the night in the hope of being able to pick up the slick again in the morning. The log of the Meigs shows that at the point

E. ASB *Preliminary Report* (p. 12/17)

the slick was found, there was a current set of 140° True at the rate of about one knot per hour. At daylight, on July 31, the Meigs attempted to relocate the slick on the theory that both the slick and the Meigs would have drifted approximately the same distance during the night. No sign of the slick was discovered and at 3:18 P.M., Manila Time, the Meigs set a course to return to the original position of the oil slick. Another search was made in that vicinity but no trace of oil was found.

Following that time, an intensive search over a large area of the ocean was conducted by the Meigs, Navy destroyers and submarines, and by Army and Navy aircraft. Other surface vessels in that part of the Pacific lent their assistance. A search of the shores and interior areas of Luzon and Mindanao, and other smaller islands of the Philippine group, was conducted by Army bombers and Navy amphibians.

During the period of the search there were no surface winds reported over 6 to 8 miles per hour, and the sea was exceptionally calm. This condition was ideal for the search for small articles such as should have been floating on the surface, had the aircraft crashed in that area, or for oil, which, in the event of a crash, would have been released in large quantities.

The search was continued until August 5, at which time the searchers felt that every possible theory as to the location of the Hawaii Clipper had been exhausted, and the search was abandoned.

An employee of a telephone company, who lives on the Island of Lahuy, is reported to have heard a large airplane flying above the clouds about 3 P.M. Manila Time, on July 29. As a result of this report, a careful aerial search was made of the Island of Lahuy and nearby areas. It was also ascertained that there were no Army, Navy or private aircraft in that vicinity on that date.

E. ASB *Preliminary Report* (p. 13/17)

Weather reports from nearby Pan American Airways Stations indicate that at the time there was an overcast, with a ceiling of about 2,500 feet, and light rain.

Lahuy is a small island lying East of and close to the Southeastern tip of Luzon, and is directly on the course from the Clipper's last reported position to Manila. At normal cruising speed the Clipper should have reached that vicinity at approximately 3:46 P. M. (Manila Time). Luzon is the largest island in the Philippine group. The Southern part of Luzon and the larger islands in the vicinity, are moderately well settled. There are, however, large areas of tropical jungles and mountain ranges which rise to an altitude exceeding 7000 feet.

The sample of oil obtained from the slick was placed in two glass jars and delivered to the U.S.S. Paul Jones. One of the samples was tested for lead content in Manila. The result was negative. The residue of this sample and the untouched sample were sent to New York for complete analysis. The samples amounted to less than 3 cc's each, which made it impossible to analyze in the usual manner; however, an investigation of the properties of these samples was conducted by chemists who were selected for their ability in this particular kind of work. At the completion of these tests they were able to announce definitely that the oil recovered from the slick was not the type of oil used in the engines of the Hawaii Clipper.

The Meigs arrived in San Francisco while the Board was in session at that point and before the laboratory tests had been made. Tests were conducted in San Francisco Bay in the presence of officers of that vessel and officials of Pan American Airways Company in an effort to identify the oil in the slick. These tests consisted of dumping samples of various kinds of oil

-15-

on the surface of the waters of the Bay under conditions that approximated as nearly as possible, those existing in the Pacific Ocean on July 30. As a result of these tests, the following facts were ascertained.

(1) - Any oil dumped on the water produces a film of the same general appearance.

(2) - The odor of high test gasoline does not remain in an oil slick more than a few minutes. This was ascertained by drawing a cloth through the slick produced from the mixture of used engine oil and 87-Octane gasoline. The film adhered to the cloth, but when drawn into the boat there was no noticeable odor of gasoline remaining, even after so short a time as five minutes. This fact is mentioned because the officers of the Meigs testified that a strong odor of gasoline was noticeable at the time they were obtaining samples of oil from the slick.

Since the officers of the Meigs had handled the sample obtained and had observed its characteristics after it had been placed in two small glass jars, the following experiments were made:

Small samples of various oils were placed in glass jars. The jars were then filled with sea water and shaken up and the result was observed for characteristic beading, odor, color and tendency to emulsify. In the opinion of the observers, none of the oil so treated appeared to duplicate the appearance or characteristics of the oil taken from the slick. It was particularly noticeable that in the experiments made with samples of new and used engine oil, the color was found to be radically different from that of the oil taken from the slick. There was aboard the Clipper 120 gallons of engine oil and a variety of other oils in small quantities; therefore, in the event of a crash, the engine oil would predominate.

E. ASB *Preliminary Report* (p. 15/17)

SUMMATION

In reviewing the evidence, it appears to the Investigating Board, that:

1. The Captain and members of the crew were qualified and physically fit.

2. Trip No. 229 progressed normally from Alameda to Guam. Inspections made at Division points along the route show that the aircraft, engines and equipment were functioning properly.

3. The aircraft and its equipment were in an airworthy condition when it departed from Guam.

4. The flight was properly dispatched from Guam to Manila in accordance with regular Company procedure.

5. Radio conditions, i.e. atmospheric conditions, were not unusual, and while communication with the plane was occasionally difficult, at no time during the trip from Alameda to Guam was communication lost. The flight from Guam west proceeded normally, and the radio equipment aboard continued to function satisfactorily until 0411 G.C.T. on July 29, 1938.

6. There was a failure of communication between the Clipper and shore stations immediately following 0411 G.C.T.

7. A wide-spread and intensive search by surface vessels and aircraft failed to disclose any evidence as to the whereabouts of the Clipper.

8. The chemical analysis of the oil sample obtained by the officers of the U. S. A.T. Meigs definitely establishes the fact that there is no connection between the oil slick discovered by them and the disappearance of the Clipper.

9. The report from the Island of Lahuy, that an aircraft was heard flying in that vicinity above the clouds on the afternoon of July 29, cannot be ignored.

E. ASB *Preliminary Report* (p. 16/17)

-17-

10. Due to the character of the terrain, the Board is not prepared to say that the aerial search of Lahuy and other areas in the islands can be considered as conclusive.

11. The reward offered by the Pan American Airways Company for information concerning the Clipper should stimulate the search by land, and may produce results.

In conclusion, it appears that the only definite facts established up to the present time, are that between 0411 and 0412 G.C.T. on July 29, there was a failure of communication between the ground and the Clipper; that communication was not thereafter reestablished; and that no trace of the aircraft has since been discovered. A number of theories have been advanced as to the possible basic cause of or reason for the disappearance of the Clipper. The Board has considered each of them. Some have not been disproved, others have been contradicted by the known facts. However, the Investigating Board feels that this report cannot properly include a discussion of conjectures unsupported by developed facts. The Board, therefore, respectfully submits this report with the thought that additional evidence may yet be discovered and the investigation completed at that time.

Robert D. Hoyt - Chairman

Phil C. Salzman - Member

W. T. Miller - Member

E. ASB *Preliminary Report* (p. 17/17)

Hoyt, Salzman and Miller were members of the original Commerce Department Board. The two auxiliary members were on the Air Safety Board.

HAWAII CLIPPER PAN AM FLT 229
NC-14714 Engines P&W 950hp each, Props Hamilton Hydromatic

Crew List: 9
Captain Leo Terletsky
1st Off Mark Walker
2nd Off George Davis
3rd Off Jose Sauceda
4th Off John Jewett ✓
Flt Engr Howard Cox
Radio Off William McCarty
Steward Ivan Parker
Mechanic Thomas Tatum

Pax- List: 6

Kenneth Kennedy (Pan Am Pacific Division Traffic Manager)
Howard French (Prosperous auto dealer from Portland, Oregon, also
                Major, Army Air Corps, Pilot, Comdr 321st Obs Sq)
Edward Wyman (Former asst to Pres Pan Am Juan Trippe, now
                VP Curtiss-Wright Aircraft Co for Export Sales.)
✓Dr. Fred Meier (Senior Scientist, National Research Council
                Wash, D.C., see previous AE Last Flight P-~~19~~ 133)
Dr. Earl Mckinley (Dean of the Medical School, George Washigton Univ
                Wash.D.C.)

Wah Sun Choy   (NJ, three resturants, 3 million dollars to HK)

Notes: AEL P-13 Guide Lines to Truk for a civilian acft, 38AE ?
        Jewett Comm AE P50 Last Flt.

EXCISED MATERIAL

Published photograph of the
christening of Hawaii Clipper

F. Passenger and Crew List (Notes)

The photograph at the bottom shows a young girl, Patricia Kennedy (no relation to passenger Kenneth Kennedy), christening *Hawaii Clipper*.

EXCISED MATERIAL

Published photograph of
Edward E. Wyman

EXCISED MATERIAL

Clipping
from Notes, p. 495
of Robert Daley's
An American Saga,
regarding the "rumors"
associated with the
loss of Hawaii Clipper,
as recounted to Daley
by Juan T. Trippe

EXCISED MATERIAL

Clipping
from Time Magazine
August 8, 1938

EXCISED MATERIAL

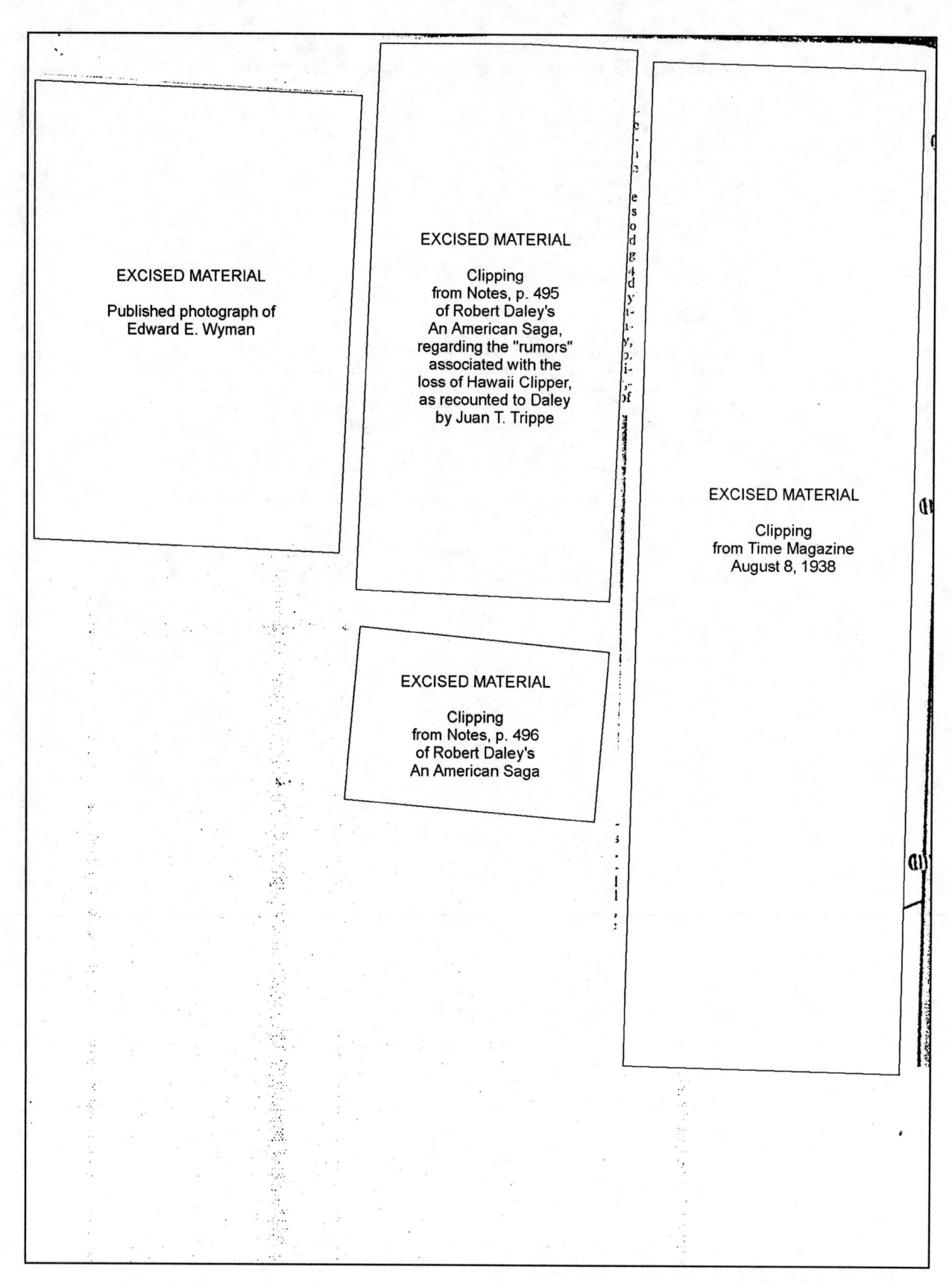

Clipping
from Notes, p. 496
of Robert Daley's
An American Saga

G. Source Material (Clippings)

Time pointedly notes a Hearst Press "question" that, with Choy's "war relief" millions aboard, "was it not a case of Japanese sabotage?"

Scource Notes: An American Saga " Pan Am Empire " Auth: Robert Daley P-496

Admiral J.T. Towers, USN who joined Pan Am as a vice president after World War II stated t
Juan Trippe, Pam Am Pres; that every Japanese Zero fighter plane examined by U.S. technici
during and after the war bore the same serial number on their magnetos............
the same serial number as the magneto on one of the four engines of the Hawaiian Clipper.

During the invasion of Saipan during WWII (the Southwest Headquarters for the Imperial
Navy) CWO W.T. Horne US Marine Corps now retired stated in the city of ~~Garapan~~ on t
island of Saipan stood the vault of the bank surrounded by reinforced concrete despite th
entire bank being levelled by shelling and bombing. Horne with a swat team of satchel
charging marines set the charge and blew the main floor off the vault. Inside they found
ten million dollars in neatly packed bundles of specially printed invasion money to be
used during the occupation of the Hawaiian Islans. Also found was three million dollars i
U.S. Gold Back paper currency of the pre -WWII era of the Pacific Invasion. The entire
contents of the vault were turned over over to a Marine Intelligence Unit headed by a
Colonel in June 1944. Thescource of the three million dollars was unknown at that time,
(Is this the money obtained from the Chines Businessman during the Hi-Jacking of the
Hawaiian Clipper in 1938 ?)

The Reluctant Admiral (Yamamoto and the Imperial Navy)
Auth: Hiroyuki Agawa Translated by John Bester Publ: Kodansha Inl, Japan 15.00-Jan 1980

P-342- Ref Natsushima Seaplane Base- Truk Atoll (Summer Island) Dublon
Battleships Yamato and Musashi 18.1 inch guns 12" armour around engine room
cannot traverse Panama Canal, Command Post's at Truk Anchorage during WWII for
Admiral Yamamoto.

P-196 - accident Report - jackson
P-496 - Daley - Pan Am
P-10 Corrugated Spars - jackson

Hull Design:
Taking a note from paper-box manufacturers, designers corrugated part
of the lower hull to acheive maximum strength of the V-bottom spars. Borrowing from
ship builders, the Martin engineers designed the Clippers with double bottoms to
help prevent sinking if the hull were damaged.

H. Source Material (Notes)

TRUK ATOLL, CAROLINE ISLANDS, THE PACIFIC;
14 November 1964.," to Dec 7,1964

(The hospital incident, Americans 1938) —Busy-Crevasse

1. Roubert Naroon, French Trukesse, Shipping Manager Truk 1937
   father, from France, a trader, mother Trukeesse native.
   " YAWATA MARU " Japanese Supply ship between Saipan & Truk 1937"
   Outcast, not freindly with Japanese authority at Dublon Base.

2. Father Nickelson on Dublon, lunch and background on Japanese
   atrocities, natives reluctant to reveal any info, still fear Japs

3. Hank Chatroop, American head of Truk Trading Company, film on
   Truk Raid.

4. Taro Mori, confirms Naroon info.

5. Sheriif Ezra Kego, student during Japanese ocuupation on Truk,
   guide to B-24 on reef.

6. Sister Ramedious, ref Sister Angelia now mother superior on
   Koror, on Saipan 1937, (no reply to ltr)

7. Mangrove swamp incident, 1939, Jap Betty Bomber, Wx accident.

8. Moses Arkana, Chief Truk Atoll in 1949, not available for intervie

9. Council of Micronesia 1964, Truk Delegates,
   Mitaro Danis- Land Title Officer
   Tosiwo Nakayama

10. Eis, living on Tol, a former Japanese, could not be located for
    interview.

11. Yosumie, employed at the court house on the island of Moen,
    a messanger for the Kempetie, (Japanese secret police during WWII
    will not see me.

12. Feik, crewman on supply ship from Truk to the Mortlock Islands,
    Ring (ace of Spades) belonged to a navigator before WWII, Jap
    Officer had it.

13. Pasition Rpt & Bearing Manila - Truk!

I. Source Material (Contacts, Truk)

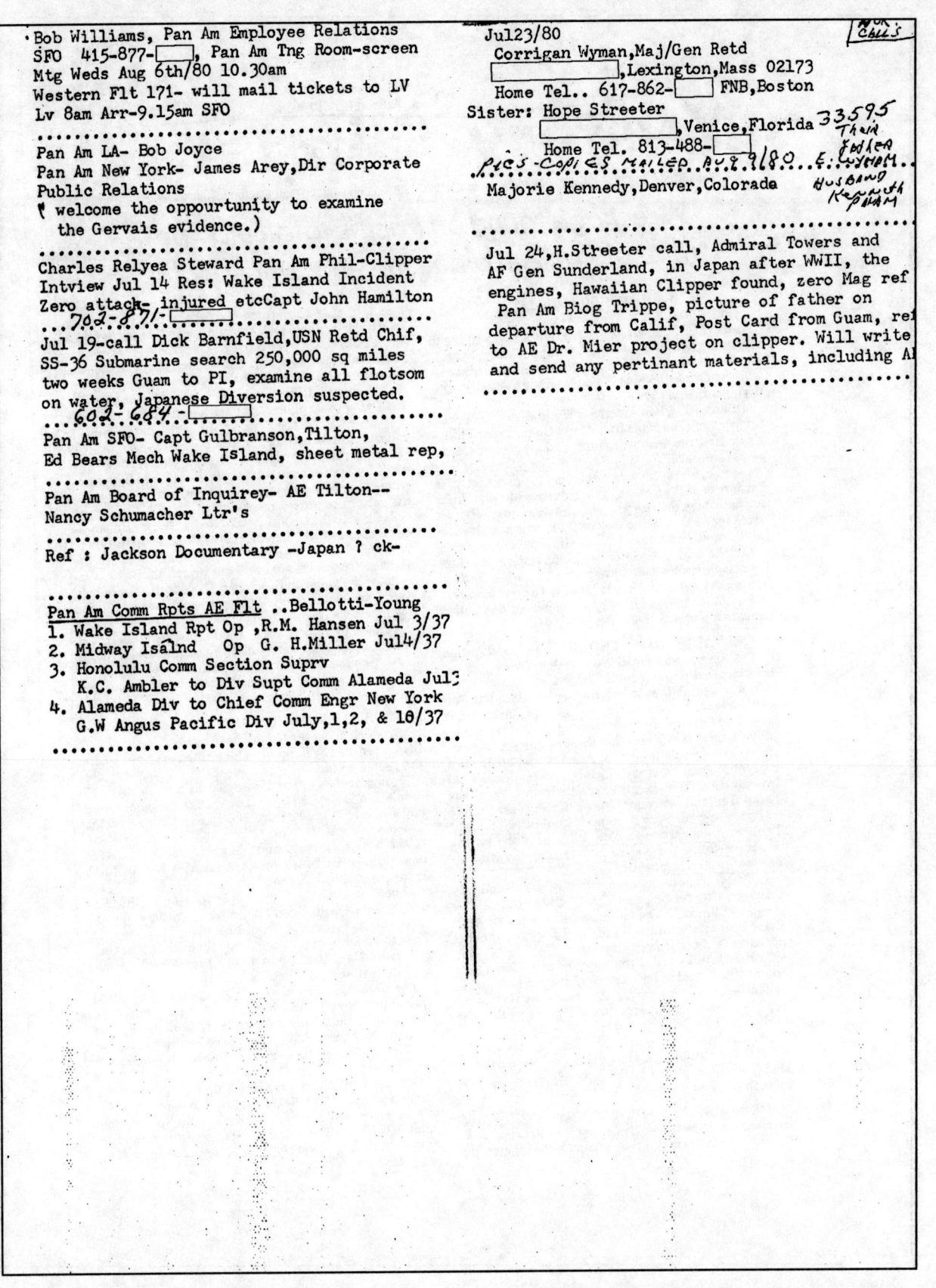

- Bob Williams, Pan Am Employee Relations
SFO 415-877-☐, Pan Am Tng Room-screen
Mtg Weds Aug 6th/80 10.30am
Western Flt 171- will mail tickets to LV
Lv 8am Arr-9.15am SFO
....
Pan Am LA- Bob Joyce
Pan Am New York- James Arey,Dir Corporate
Public Relations
(welcome the oppourtunity to examine
   the Gervais evidence.)
....
Charles Relyea Steward Pan Am Phil-Clipper
Intview Jul 14 Res: Wake Island Incident
Zero attack- injured etcCapt John Hamilton
   702-87/-☐
Jul 19-call Dick Barnfield,USN Retd Chif,
SS-36 Submarine search 250,000 sq miles
two weeks Guam to PI, examine all flotsom
on water, Japanese Diversion suspected.
   602-684-☐
Pan Am SFO- Capt Gulbranson,Tilton,
Ed Bears Mech Wake Island, sheet metal rep,
....
Pan Am Board of Inquirey- AE Tilton--
Nancy Schumacher Ltr's

Ref : Jackson Documentary -Japan ? ck-

....
Pan Am Comm Rpts AE Flt ..Bellotti-Young
1. Wake Island Rpt Op ,R.M. Hansen Jul 3/37
2. Midway Isalnd Op G. H.Miller Jul4/37
3. Honolulu Comm Section Suprv
   K.C. Ambler to Div Supt Comm Alameda Jul3
4. Alameda Div to Chief Comm Engr New York
   G.W Angus Pacific Div July,1,2, & 10/37
....

Jul23/80
Corrigan Wyman,Maj/Gen Retd
☐,Lexington,Mass 02173
Home Tel.. 617-862-☐ FNB,Boston
Sister: Hope Streeter
☐,Venice,Florida 33595
Home Tel. 813-488-☐ *Their Eddie*
Pics-copies mailed Aug 8/80 E. Lynum
Majorie Kennedy,Denver,Colorado *Husband Kenneth Guam*

....
Jul 24,H.Streeter call, Admiral Towers and
AF Gen Sunderland, in Japan after WWII, the
engines, Hawaiian Clipper found, zero Mag ref
Pan Am Biog Trippe, picture of father on
departure from Calif, Post Card from Guam, re
to AE Dr. Mier project on clipper. Will write
and send any pertinent materials, including A
....

J. Source Material (Contacts, U.S.)

Note: street addresses and telephone numbers, inappropriate after 20 years, have been defaced to protect present residents and phone users.

"New York Times & Los Angeles Times July 6, 1980"

Mystery Clipper

Wreckage found on Pacific atoll may be that of Pan Am plane missing since 1938

By PATRICK ARNOLD

LAS VEGAS, Nev. (AP) — Joe Gervais was trying to solve the greatest mystery in aviation history, the disappearance of Amelia Earhart, when he stumbled across what he now believes is the answer to an equally mystifying airplane puzzle.

In July 1938, Pan American World Airways' Hawaii Clipper vanished while flying from Guam to the Philippines. No trace was ever found of the plane or the 15 people on board.

Gervais, who spent a decade trying to prove that Miss Earhart is still alive, says that during his search for evidence more than 15 years ago he was shown aircraft wreckage on the Pacific atoll of Truk, wreckage he now thinks could be that of the Hawaii Clipper.

And, he said in an interview, information indicates the disappearance may have been the world's first hijacking. A local islander told him the 15 crew members and passengers were killed by the Japanese and buried beneath a hospital on the island.

GERVAIS SAID HE THOUGHT an investigation by Pan Am was warranted to determine if the wreckage is in fact that of the Hawaii Clipper.

"I would think that to look further in this ... Pan Am would have to become interested," he said. "After all, it was a Pan Am airplane, their passengers, their responsibility. If there is evidence that those people are there on that island, I would think that Pan Am ought to make a request to Trust Territory officials ... to conduct a local investigation. That is, go where the remnants of the hospital are and determine if those 15 people are there."

James Arey, director of corporate public relations at Pan Am headquarters in New York, said the airline would "welcome the opportunity to examine the evidence that Mr. Gervais has."

Gervais said he realized what he may have found only this year, after reading a new book, "China Clipper" by Ronald W. Jackson. The book is a history of Pan Am's flying boat service in the Pacific and discusses the disappearance of the Hawaii Clipper.

Gervais, a retired Air Force pilot and now a Las Vegas truant officer, said he had flown to Truk — a formidable Japanese naval base during the war — in November 1964 to investigate the wreckage of a plane he thought could have been Miss Earhart's.

Gervais based his search for Miss Earhart on the belief the famous flyer and navigator Fred Noonan were shot down and captured by the Japanese while on a spy mission in 1937.

GERVAIS SAID THE WRECKAGE a Trukese native showed him was obviously not the Earhart plane. Gervais now believes it is the remains of the clipper. He was told that 15 persons aboard the flying boat were executed by the Japanese and was shown where the bodies allegedly were buried.

Gervais said he didn't think much about it then because he was unfamiliar with the missing clipper story.

The Hawaii Clipper, one of several four-engine Pan Am flying boats plying the Pacific in the late 1930s, was last heard from about noon, July 29, 1938, when it radioed its position and weather report.

No further messages were received, setting off a week-long sea search. A trace of oil was found in the area where the plane was last heard from, but an analysis of the substance proved inconclusive.

Jackson, in his book, theorizes that two Japanese naval officers from Saipan, about 100 miles from Guam, sneaked aboard the clipper when it anchored for the night at Guam. He suggests the two officers took command of the plane and ordered it flown to Japanese-held Koror, an island about 400 miles southwest of Guam.

Gervais disagrees, saying it was more likely the plane was flown to Truk, part of the same island chain, because the Japanese were building a seaplane base there.

Jackson has told Gervais he hadn't known about the Truk wreckage and expressed interest in pursuing the new theory, according to Gervais.

GERVAIS SAID HE "TOOK a few pictures" of the craft, which appeared to have been cut into sections, and then was asked by his guide if he wanted to know "where the people were."

"And I said, 'I'm not interested in a plane with 15 people. I'm interested in a plane with two people, a man and a woman.' And he said, 'No, 15 Americans, and the Japanese executed them.'"

The bodies, Gervais said his guide told him, lie beneath the crumbling foundation of a bombed-out Japanese hospital on the atoll.

"I don't know how many people were on it (the clipper)," Gervais said. "I don't know whether there were 40 people on it, three or 10, and I had forgotten about it. Then this book came out and I thought back at that time and said, 'Hey, that's how many people he told me. Fifteen. Not 45 or 30, but 15."

The book also describes a plane passenger named Wah Sun Choy, a New Jersey restaurant owner and chairman of the Chinese War Relief Committee who reportedly was carrying as much as $3 million raised in the United States to help the Chinese in their war against Japan.

Gervais said that when on Saipan pursuing his Earhart investigation, he talked to a retired Marine who told him that while the military was searching a bank vault in Garapan, the island's largest city, they found $3 million in old U.S. "goldback" currency in use during the pre-war years.

K. Associated Press Story (7/6/80)

Reprinted by permission of the Associated Press

The death knell for Gervais' 1980 investigation was sounded in both leads for this AP release, which focused on the wreckage found at Truk.

EXCISED MATERIAL

Magazine advertisement
for the Continental Hotel
on Moen Island, Truk,
August, 1980

L. Continental Hotel Advertisement   Magazine advertisement for accommodations at Moen Island, Truk (now Chuuk State, Federated States of Micronesia) as of about August, 1980).

TRUST TERRITORY OF THE PACIFIC ISLANDS
OFFICE OF THE HIGH COMMISSIONER
SAIPAN, MARIANA ISLANDS
96950

COMMERCIAL
CABLE ADDRESS
HICOTT SAIPAN

November 9, 1964

Mr. Joseph A. L. Gervais
1905 Theresa Avenue
Las Vegas, Nevada

Dear Mr. Gervais:

Enclosed is a validated Trust Territory Visitor Permit which will authorize you to visit the Truk District for the purpose of completing historical research of the former Japanese naval bases located on Moen, Dublon and Etan Islands. We have also included a possible side trip to Saipan as indicated in the last paragraph of your letter.

We thank you for the most informative enclosures which you included with your application. They are being returned per your request.

If we may be of any assistance, please do not hesitate to call upon us.

Sincerely,

R. A. Johnson
Immigration Administrator

Enclosures

M. Visitor Permit (Cover Letter)

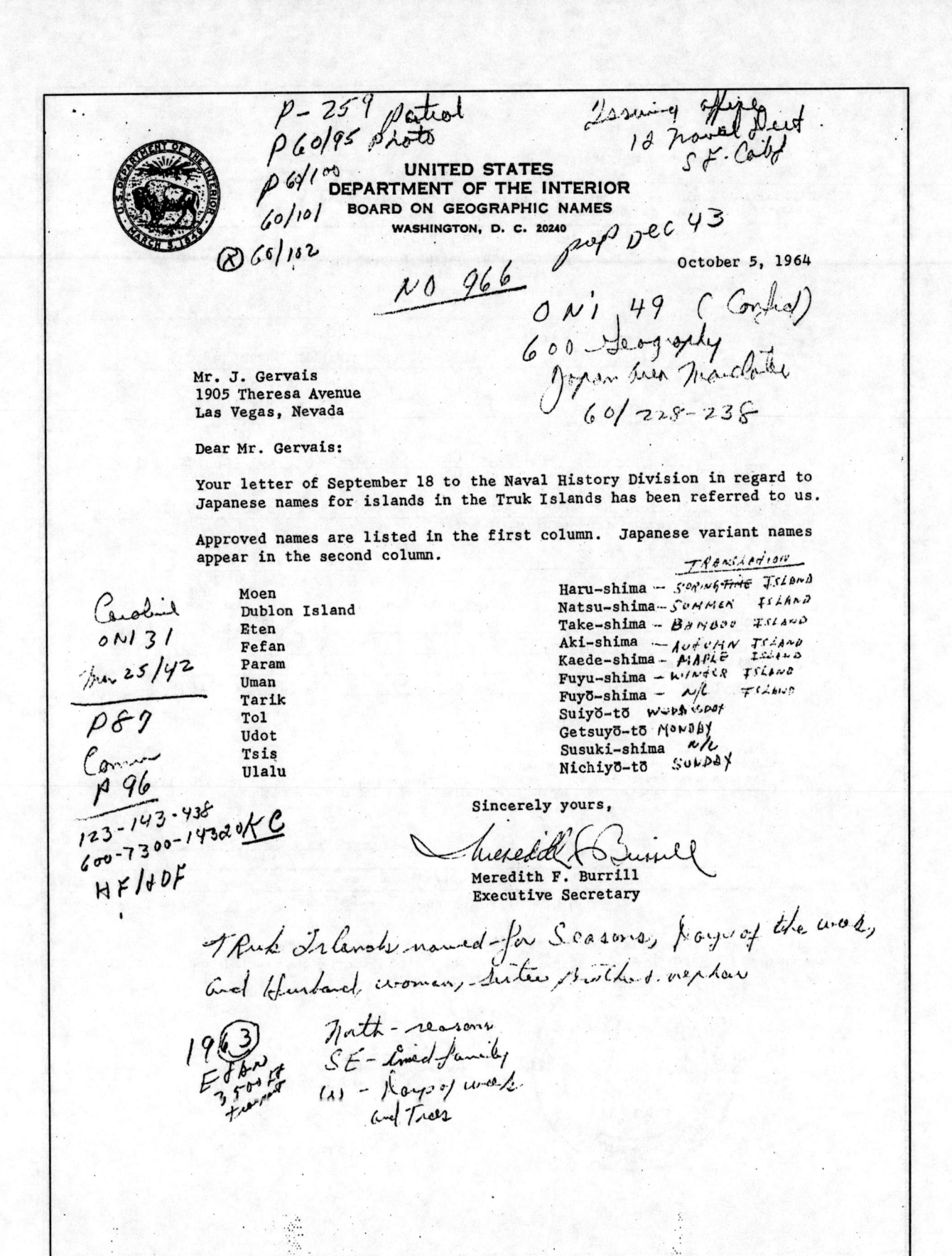

**UNITED STATES
DEPARTMENT OF THE INTERIOR
BOARD ON GEOGRAPHIC NAMES
WASHINGTON, D. C. 20240**

October 5, 1964

Mr. J. Gervais
1905 Theresa Avenue
Las Vegas, Nevada

Dear Mr. Gervais:

Your letter of September 18 to the Naval History Division in regard to Japanese names for islands in the Truk Islands has been referred to us.

Approved names are listed in the first column. Japanese variant names appear in the second column.

	TRANSLATION
Moen	Haru-shima — SPRINGTIME ISLAND
Dublon Island	Natsu-shima — SUMMER ISLAND
Eten	Take-shima — BAMBOO ISLAND
Fefan	Aki-shima — AUTUMN ISLAND
Param	Kaede-shima — MAPLE ISLAND
Uman	Fuyu-shima — WINTER ISLAND
Tarik	Fuyō-shima — N/L
Tol	Suiyō-tō — WEDNESDAY
Udot	Getsuyō-tō — MONDAY
Tsis	Susuki-shima — N/L
Ulalu	Nichiyō-tō — SUNDAY

Sincerely yours,

Meredith F. Burrill
Executive Secretary

N. Island Names (Truk Atoll)

TT FORM NO. 97 (Revised July 1, 1963)

Read carefully conditions governing entry into the Trust Territory printed on the reverse side hereof before preparing and submitting application.

APPLICATION AND PERMIT TO VISIT
THE TRUST TERRITORY OF THE PACIFIC ISLANDS
(Submit in duplicate)

November 2, 1964
DATE

1. (MR.) Joseph A.L. Gervais — 1905 Theresa Ave, Las Vegas, Nevada, USA
   NAME — MAILING ADDRESS

2. Same as mailing address — U.S.A.
   PERMANENT HOME ADDRESS — NATIONALITY

3. E 732374 — Dept of State, Los Angeles, Calif Oct 19, 1964 — Los Angeles, Calif.
   PASSPORT NUMBER — ISSUED BY — WHERE

4. Complete Historical Research of Japans formidable Navy Base during WWll see covering letter for details
   PURPOSE OF VISIT

5. May 19, 1924 — Tyngsboro, Massachusetts.
   DATE OF BIRTH — PLACE OF BIRTH

6.
AREAS TO BE VISITED	LENGTH OF VISIT	DATE OF ENTRY AND CARRIER
A. Truk District	one week	Will Arrive Guam Approx- between 16 & 25 Nov 64
B. (To include Moen, Dublon,		
C. & Etan Islands.)		
D. Saipan	two days	TT aircraft

I certify that the facts hereinabove set forth are true and correct to the best of my knowledge and belief; and it is fully understood that throughout the period of my visit, I am subject to all of the rules, regulations, and laws of the Trust Territory. If permit is withdrawn for any reason while I am in the Trust Territory I agree to leave said Territory by the first available transportation at my own expense.

Joseph A.L. Gervais
SIGNATURE

DO NOT WRITE BELOW—FOR OFFICIAL USE ONLY

VISITORS PERMIT NUMBER ––2946––

Authority is herewith granted, this 9th day of November 1964, for the entry of Mr. Joseph A.L. GERVAIS into the Trust Territory of the Pacific Islands, to remain therein for a period of two (2) weeks from date of entry.

[Stamp: ADMITTED NOV 21 1964 SAIPAN, M.I. TRUST TERRITORY PACIFIC ISLANDS MARIANA ISLANDS DISTRICT]

R. A. JOHNSON — ISSUING OFFICER
Immigration Administrator

O. Visitor Permit 2946 (Trust Terr.)

Note the stamp certifying entry into the Marianas Islands District (Saipan), on November 21, 1964. Red tape and transport limitations isolated Truk.

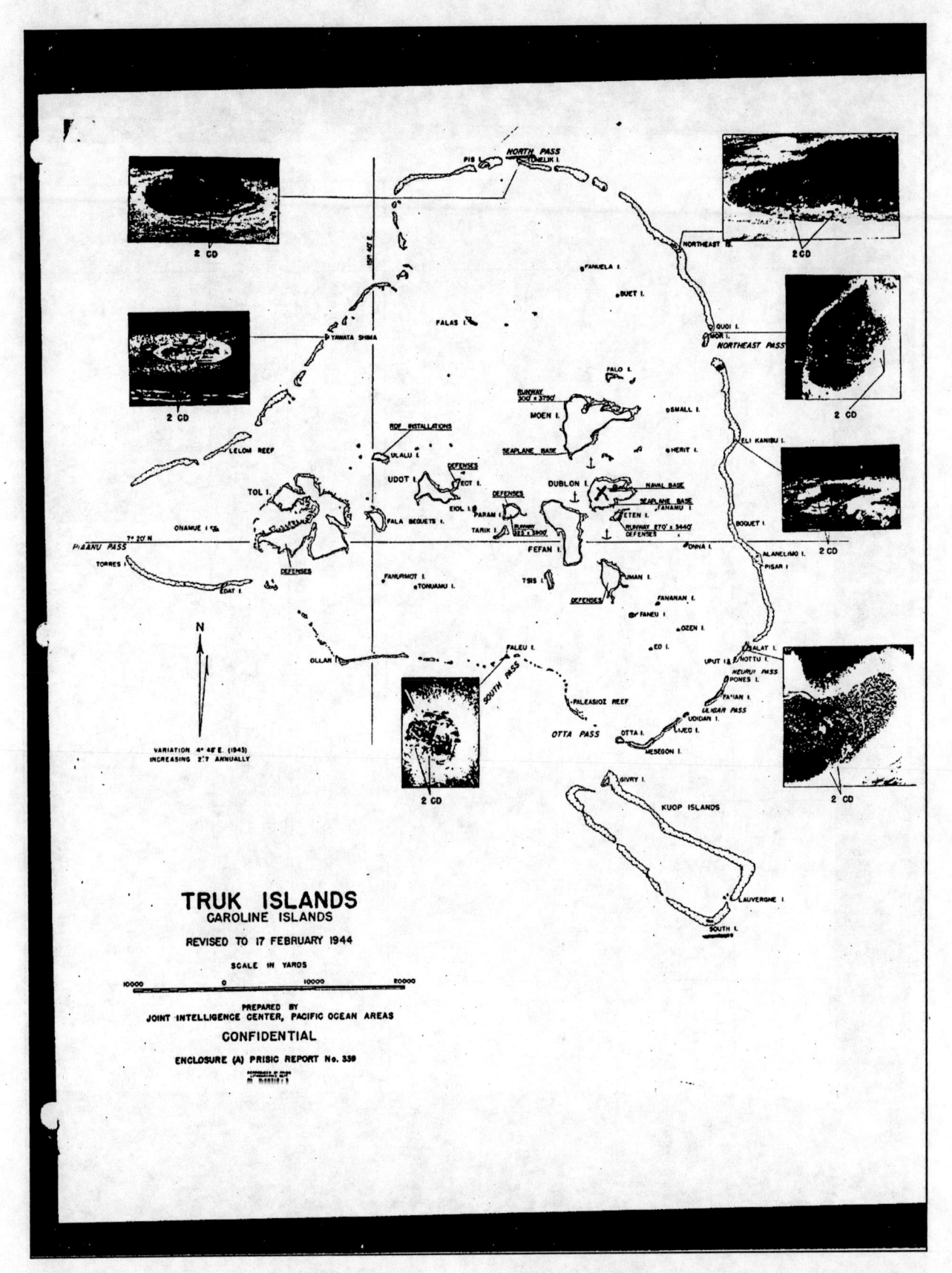

P. Truk Atoll (WWII JICPOA Map)

Military Intelligence map of Truk, "revised to 17 February 1944," immediately after Mitscher's air assault on Truk, prior to the Eniwetok invasion.

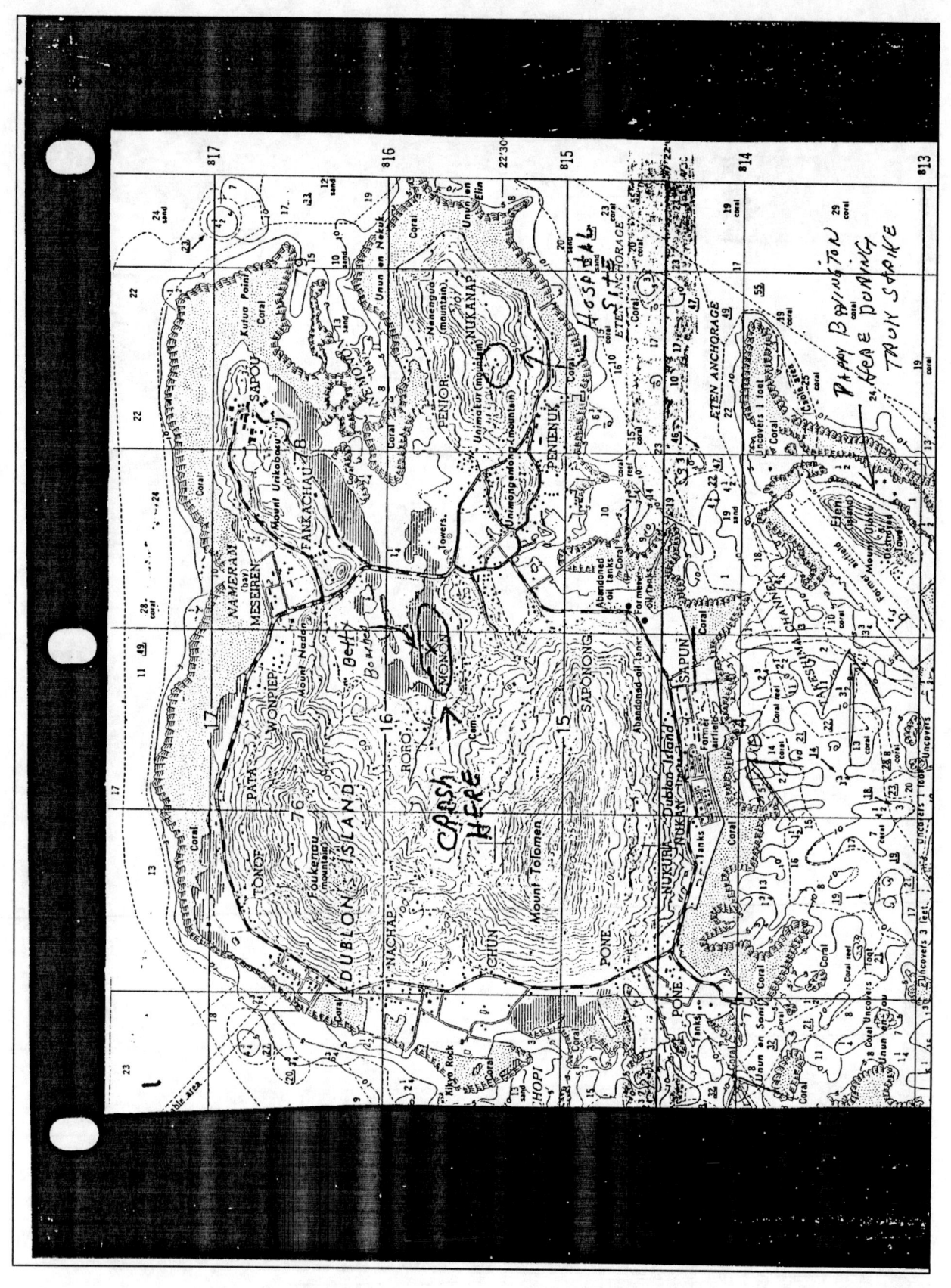

Q. Dublon Island (USCGS Chart)

Comments were written on the chart by Gervais; "CRASH HERE" refers to a WWII *Betty* bomber, which Joe found in the mangrove swamp.

Nov 16, 1964: The stone post on the left marks the entry way to the Japanese Navy hospital located on the island of Dublon within the Truk Atoll of the Caroline Islands. The Japanese name of this island during WWII was Natsushima, meaning " Summer Island."The entire hospital was destroyed during the air strike on Truk in February 1944. The jungle growth has almost entirely reclaimed the original site except the cement slab foundation which is intact. Sections of the slab can be seen at the base of the palm tree. Beneath this slab is where the 15 personnel of the Hawaiian Clipper were placed, on the north side, and that section refferred to as the rear of the hospital.

R. Hospital Slab (Photo and Text)

After Gervais broke the story, in 1980, the site of the slab became a popular attraction with the many Japanese tourists who visited Truk.

Nov 16, 1964: A Japanese Navy personnel transport used within the Truk Atoll to move personnel from Warships to shore facilities during WWII. The aluminum pre-fab covering over the wooden barge was made from sections of the hull of the Hawaiian Clipper. Purpose of the cover was to keep high ranking Japanese Naval Officers dry during rainsqualls. The barge was reffered to by Trukee's natives, "The Admiral's Barge."

S. Personnel Barge (Photo and Text)

The round portholes on sections of this "barge" eliminate *Hawaii Clipper* (rectangular windows) as a source for any of the visible components.

Nov 16, 1964: A part of the mid-section located aft of the flight deck of the Hawaiian Clipper. The crew compartment walkway door is visible. Behind the V-bottom spar on the right a section of the left wing is visible pertruding above the water. None of the four engines from the clipper were located within the Truk Atoll.

T. Aircraft Wreck (Photo and Text)   Gervais' mistaken identification of this wreckage, as being the *Hawaii Clipper*, provided authorities with cause to reject the Nauroon story as well.

> Nov 16,1964: The nose section of the Hawaiian Clipper located in
> shallow water on the southern tip of Moen Island in the Truk Atoll
> of the Caroline Islands. Windows from the flight deck have been removed.
> Borrowing from ship builders, the Martin engineers designed the Clippers
> V-bottom spars with corrugated metal and double bottom hulls to prevent
> sinking if damaged on landing.

U. Aircraft Wreck (Photo and Text)

Unidentified aircraft cockpit on Moen Island, at Truk. Even in its then-present state, this wreck is definitely not the "bridge" of *Hawaii Clipper*

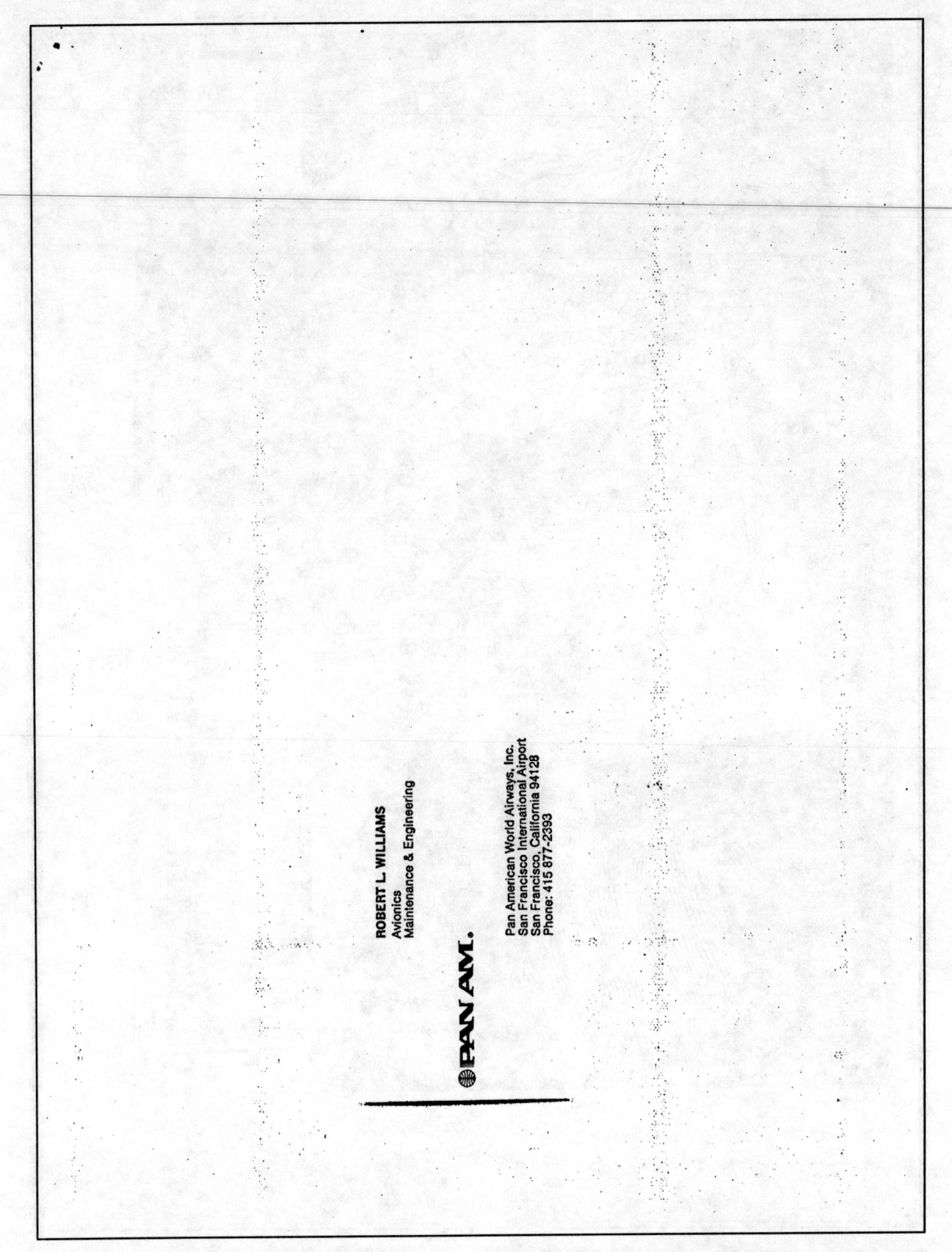

ROBERT L WILLIAMS
Avionics
Maintenance & Engineering

PAN AM.

Pan American World Airways, Inc.
San Francisco International Airport
San Francisco, California 94128
Phone: 415 877-2393

V. Pan Am Business Card (Contact)

Business Card of Robert Williams, who helped Joe Gervais set up the August 6, 1980 meeting with PAA retirees at Pan Am, San Francisco.

W. Pan Am Crew Conference (Photo)

Joe Gervais (striped shirt) meeting, on August 6, 1980, with PAA retirees (pilots at table with Joe) in the Pan Am training room, in San Francisco.

3. EARHART AT MILI: Resolving AE's 157-337 Line of Position—the Reference Point

INTRODUCTION

On the morning that she transmitted her last two radio messages and disappeared, Amelia Earhart wasn't lost. She knew exactly where she was and, what's more, she revealed her location to anyone who had the wit—and the chart—with which to decipher those last two messages, which actually formed two parts of a single position estimate. However, the second of these messages, or parts, was too cryptic to be immediately understood, even by those who may have been "in the know."

Because of the sensitive nature of Earhart's geographic position that morning, it is quite understandable that she felt a need to reveal that location discreetly and without fanfare. But, because Earhart had no apparent reason for being where she actually was, her resort to cryptic word games was strangely out of place. Considering the depth of her crisis, she was playing a most dangerous game and risked missing the mark (as she did miss it) with the only men who could have rescued her—and changed the course of history: the officers and crew of the Cutter *Itasca*.

Lt. Cdr. Frank Kenner had heard Earhart's last broadcasts in the radio room aboard the U.S. Coast Guard Cutter *Itasca*, which had been awaiting her arrival at Howland Island. In a personal letter, written in Honolulu on August 10, apparently to his mother-in-law, Kenner quietly detailed his impressions of Earhart's involvement with *Itasca*: "...she had only herself to blame...she was too sure of herself, and too casual. She devoted no effort to the details at all. When it was too late and she was going down *she hollered for our aid* but that was too late...*She never gave us any of her positions*...I heard her last broadcasts myself. She realized too late that she was in trouble *and then she went to pieces*. Her voice plainly indicated that fact, by the desperate note in her transmissions. She asked us to do the impossible, knowing ahead of time that we could not furnish her with the service that she wanted. She clearly indicated throughout the flight that she was not familiar with her radio equipment. If she had only answered our messages earlier in the flight *we might have had some idea where to look for her*, and might have been able to save her...we really cannot find all the answers [Italics added by the author]."

Apparently, Earhart had more to say, during her final broadcasts, than is to be found in the *Itasca*'s radio logs. The radiomen logged only what seemed to be the "pertinent data" (as O'Hare's 8:56am log entry specifically states), and could not be concerned with her desperate remarks. The officers, on the other hand, having no records to keep, were too engrossed, perhaps, in the growing hysteria of a woman about to die, to appreciate fully the few cryptic bits of "pertinent data" that Earhart transmitted. Perhaps understandably, therefore, neither the officers nor radiomen deciphered her final "encrypted" message. They would hardly have believed it, even if they had deciphered it, since Mili Atoll, in the Japanese-ruled Marshall Islands was 600 nautical miles from Howland Island and the Coast Guard Cutter *Itasca*. And Mili Atoll is just where Earhart's *Electra* went down, at about 9:05am, *Itasca* time, on July 2, 1937.

This is nothing new, of course. Earhart has been privately—and secretly—linked to the Marshall Islands since a few days after her disappearance. Many of these links have been known to the reading public since the 1960's and known among serious historians since much earlier. What is new, here, is the proof that *Earhart, herself, provided the first of these links*—just before she ditched!

POINTS OF CONCERN

Although the proof of this link, or "bridge to the Marshalls," is fairly self-contained, there are a few points of concern that must be considered, before examining the evidence.

A. First, basic consideration must be given to the topic of radio wave propagation, in order to correct a false assumption made by *Itasca*'s Chief Radioman, Leo Bellarts, on the morning of Earhart's disappearance. He had noted a rising strength in Earhart's radio signals from S1 at 4:43am, to S3 at 6:15am, to S4 at 6:45am and, finally, to a maximum of S5, between 7:42am and her last message at 8:43am (or 8:55am). Perhaps because he had no reason to assume that Earhart could be anywhere but generally on her course to Howland Island, Bellarts also assumed that her rising signal strength was derived from short-range *Ground Waves*, as she closed in on the *Itasca*, at Howland Island.

Those having experience in high frequency, or "short wave," communications, however, know only too well that dawn brings the most radical of daily changes in radio wave propagation. The author had a year of such experience, from December of 1962 to late November of 1963, at the U.S. Coast Guard Loran Transmitting Station, Johnston Island, *where rising signal strengths at dawn were a daily occurrence*, on a night frequency of 4150kHz, from stations 700 miles away.

Earhart's assigned radio frequencies lay in the lower High Frequency (HF) band and were standard, in 1937, for voice (AM) use by "itinerant and military aircraft:" 3105kHz for night use, and 6210kHz for daytime use. There are three modes of propagation for HF radio waves: Line of Sight, with a range that is limited by the altitude of both the receiver and the transmitter; Ground Waves, which travel across (and within) the earth's surface with a range of about 60 miles over land and about 100 miles over the sea (independent of frequency at sea); and Sky Waves, which, are reflected from (actually *refracted within*) active layers in the ionosphere and returned to earth at great distances from the point of transmission, depending upon the frequency, the angle of incidence at which the waves enter a layer (to permit refraction) and the time of day (for any one particular day).

Although the angle of upward radiation can be roughly directed by an antenna's design, radio waves from long-wire antennas radiate outward uniformly, from just above the horizon to the zenith. For a given latitude, there will be a maximum usable frequency, above which, signals at a high angle (nearer the zenith) will not be refracted, but will pass through the layer and out into space. As these signals are not returned to earth, there will be a *skip zone* in the Sky Wave return. Reception, then, will be by Line of Sight, in sight of the transmitting antenna; by Ground Wave, out 60 to 100 miles; *then a skip zone of no reception*, from the outer radius of the Ground Waves to the first Sky Wave return area; then reception for hundreds of miles.

During the day, the sun ionizes the lower D, E and F1 layers, and the E layer, at only about 60 miles altitude, becomes the major active layer. A day frequency of 6210kHz will "skip," out to about 100 miles, with reception beyond the skip, out to 600 miles.

At night, when the lower layers cannot be ionized by the sun, Sky Wave "reflection" occurs at 180 miles altitude, in the highest layer, the F2, which is independent of solar altitude (and merges, at night, with the F1). A night frequency of 3105kHz will provide a "skip" to about 200 miles, beyond which reception will extend outward to 1500 miles (for a single "hop"), while a day frequency of 6210kHz will skip, at night, to about 500 miles, with reception to 2000 miles. Late on June 30, and several hours later, at 5:45am, July 1, *Itasca* had worked Harry Balfour at Lae on *12000kHz* (25 meters) at 2500 miles.

Just after dawn, however, as ionization of the lower layers (D, E, F1) begins, the layers begin dropping to lower altitudes and radio wave propagation exhibits the most radical changes of the day. The reflection and skip zones of both upper and lower layers move across the surface, altering reception. On a suitable night frequency, noise and signals increase in strength to a maximum S5, but signals come through crisp and clear until an hour past dawn, when the night frequency will suddenly "drop out" without warning.

The author spent the better part of 1963 observing this phenomenon every morning, before shifting from 4150kHz to 9630kHz, and on rare occasions, when Loran problems interfered with a shift to the day frequency, S5 communications invariably dropped out after dawn, with astonishing suddenness.

The high strength of Earhart's signals, at 8:43am on July 2, therefore, *cannot be used to prove her proximity to the Itasca*. In fact, had she been, as *Itasca* assumed, only 40 miles to the northwest, then her promised repeat of the 8:43am message, on 6210kHz, *would have been heard at the same strength as on 3105kHz*, because Ground Waves are independent of frequency at sea. And yet her 6210kHz signals *were never heard by Itasca at any time*, although they had been heard on the previous afternoon, by Balfour, in Lae. As to her 3105kHz night frequency, she was never *un*heard by *Itasca*, from 2:48am to 8:43am, for more than 81 minutes. Had she been near Howland at 8:43am, her signals would have been erratic or unheard from 6:14am (reportedly at "200 miles out") to 8:00am as her skip zone passed over *Itasca*.

B. Second, while the author has no doubt as to the fine service provided to Earhart by the officers and crew of *Itasca*, there were a few elements relating to radio communications, over which *Itasca* had no control, but which led to a great deal confusion. One of these is worth a chapter, in itself, and cannot easily be presented within the scope of this section of the Appendix, but a few of these elements are critical to a discussion of Earhart's final broadcast revelation, as to her final location.

Radio communication in the late 1930's, and especially at sea, was primarily a matter of wireless telegraphy (using the International Morse Code and the "shorthand" of three-character "Q" and "Z" signals)—*not voice*. Operators were trained to copy information as a stream of characters, at a typical rate of fifteen words per minute. This information could be in plain language, in abbreviations, in five-character code groups or, typically, in "Q" or "Z" signals. But, no matter what form the *information* took—as telegraphic "data," it all came in at a standard, rhythmic rate of fifteen words per minute.

It may be stated, with confidence, that *Itasca* had good reason to believe, before the flight, that all radio communication with Earhart would be at the standard wireless telegraph rate of fifteen words per minute. It does not appear that *Itasca's* radiomen ever realized that, as things actually turned out, wireless telegraphy was totally useless in the case of Earhart's last flight, but they may have been stunned to discover just how difficult it was to copy *voice* transmissions under the stress which attended this high profile operation.

Voice comes through, not as a simple stream of *characters*, at fifteen words per minute, but as a stream of *words*, at a higher rate. In Earhart's case, it came through as a veritable torrent of words, at a *rate* perhaps as high as seventy-five words, or more, per minute.

In fact, on page 35 of the *Itasca's* report, "Radio Transcripts, Earhart Flight," note is made of Earhart's exclusive use of voice and of the resulting difficulties: "The transcript of the radio logs from 0200 until 0930 is necessarily *not complete* due to the rapidity of events and also due to the Earhart exclusive use of voice, *only partially received* [italics added by the author]." On page 43, note is made of Earhart's failing stability as the long flight neared its end: "Toward the end Earhart talked *so rapidly as to be almost incoherent*." But, of course, it was 'toward the end' that she sent her most important messages, having to do with her location. In addition, it should be noted that, except for the Chief Radioman, Leo Bellarts, and the second class radioman, Frank Cipriani (who was on temporary duty from the *Taney* and assigned to RDF duty on Howland Island), all three of the duty radiomen on *Itasca* held the rate of Radioman Third Class. On page 35 of "Radio Transcripts," it is clearly noted that, "the ITASCA's radio personnel in the beginning were inexperienced." The report also notes, and rightly so, that "the operators were interested and proved capable in the heavy load which they carried," but they did lack the essential experience for the special task which they were expected to perform.

Having neither training nor experience as stenographers, *Itasca*'s radiomen were not able to copy the *words* which they heard, in the manner in which they copied telegraphic signals, that is, character by character. They had to absorb an entire phrase or sentence, evaluate its meaning, paraphrase it briefly and copy it on the typewriter, using as many of the original words as they recalled, while, at the same time, allowing the next phrase to begin registering in their minds. With regard to Earhart's last messages, *Itasca* copied key *words* which Earhart had transmitted, but, even with so many witnesses present in the radio room that morning, the meaning of the *phrases* which she sent were misunderstood and, thus, incorrectly copied and logged.

C. Third, the aftermath of Earhart's loss requires some preliminary attention, as well. In the first place, the *Itasca*'s report, "Radio Transcripts, Earhart Flight," was assembled, during the search for Earhart, using radio log *transcripts*, not the original log sheets, as the primary sources for the report. This led to the first published errors regarding Earhart's final messages. Correction of these errors was, therefore, not possible until a few of the original log pages turned up in 1975:

When Chief Radioman Leo Bellarts left the *Itasca*, he took with him a batch of original radio messages and four of the *original* log pages. These pages were from the special radio position that had been set up to log the Earhart communications. During the last contact period, from 8:43am to 9:07am, Bill Galten, Rm 3C had that watch, while Tom O'Hare, Rm 3C, had the watch at the ship's primary radio position (and Frank Cipriani, Rm 2C, had the special High Frequency Radio Direction Finder watch on Howland Island, well away from the steel hull of the *Itasca*). When Tom O'Hare left *Itasca*, he also reportedly took a batch of original radio messages and may have taken the log from the ship's main radio position, as well. With so much missing, the truth has had to wait.

O'Hare's collection of original material may still remain in private hands, but Bellarts' material was bequeathed, by his heirs, to the National Archives, in 1975. Included in this collection of letters and clippings are the four original log pages (one blank) from the special Earhart radio position. One of these pages, for the time from 8:05am to 10:39am, which includes Earhart's last recorded radio transmission, has been reproduced for this section, along with a blow-up of a critical detail on the sheet—a detail which is not, of course, revealed in the transcript. As is the case with most records of communication, especially Earhart's final message, there is often far more to the *meaning* of things, than the words, alone, will convey.

In the immediate aftermath of Earhart's loss, there was one very puzzling development. At 10:15am, as *Itasca* completed her hurried preparations to depart Howland Island and commence a search for Earhart, the ship sent a message to the Coast Guard San Francisco

Division (NMC) and the Hawaiian Section. The message stated, in part, that Earhart had, at, "0843 REPORTED LINE OF POSITION 157 DASH 337 BUT NO REFERENCE POINT PRESUME HOWLAND PERIOD".

But Earhart's last message, *as entered in the radio log at 8:43am*, includes not only the line of position, but also the *north and south comment*, which, up to the present day, has been incorrectly interpreted as, "we are running north and south on the line," presumably, of course, a reference to the 157-337 reciprocal *line of position*.

The "comment" was left out of this 10:15am message, even though the sheet on which it was logged was still active at that time, and would be, until 10:39am. A message sent at 2:02pm recounted the 8:43am message, just as the 10:15am message had done, *and also omitted the north and south comment*.

However, a message sent at 8:15pm stated that, "ITASCA CONTACTED EARHART TO RECEIVE INCOMPLETE MESSAGES ON AGREED SCHEDULES FROM 0248 TO 0855 THIS MORNING PERIOD". It then went on to state that, "HAVE HEARD NO SIGNALS FROM EARHART SINCE 0855 THIS MORNING WHEN SHE GAVE ITASCA A LINE OF POSITION [8:43am] BELIEVED TO MEAN RADIO BEARING AND STATED SHE WAS RUNNING NORTH AND SOUTH PERIOD [8:55am; under-scoring added by the author]".

This was repeated in a message to the Navy at 2:05am, *but "0855" was dropped from all further reports.* It is unclear as to why no further reference was made, as to the 8:55am transmission, but it may have been related to the great length of that last transmission, the lack of "pertinent data" in it and the serious questions that would have been raised. Still, the *north and south* comment was *firmly linked* to a last transmission on the morning of July 2—not at 8:43am—but at 8:55am.

SUMMARY OF THE POINTS

In summarizing the introduction, it has been established that:

a. Earhart's high signal strength (S5) on her night frequency of 3105kHz, from 7:58am until her last transmission, after 8:43am, in no way proves that she was within ground wave range (100 miles) of *Itasca*, (at Howland Island) during that short period. On the contrary, *Itasca*'s inability to hear the daytime frequency of 6210kHz after 8:43am, when Earhart switched to that frequency, indicates that *she most certainly was not within ground wave range during the final hour.*

b. The inexperience of *Itasca*'s radiomen in copying voice broadcasts, coupled with Earhart's reportedly agitated state, rapid speech and apparent incoherence, led to an admission, on the part of *Itasca*'s own officers, that the messages received from Earhart were, at the least, "not complete" and "partially received." This opens the door to consideration of the messages as having been *misunderstood*, as well.

c. *Itasca*'s amendment of Earhart data, in outgoing messages sent between 8:15pm and 2:05am (i.e., adding the *north and south* comment and revising the receipt of her final message from 8:43am to *8:55am*), clearly confirms that the radio log for the Earhart radio position had, indeed, been amended, after the fact, to include the *north and south* comment. It also confirms that there was, indeed, a broadcast at 8:55am, in which Earhart "hollered for...aid," but which was not logged, except for the *north and south* comment. The *temporary* status, in the outgoing reports, of an 8:55am time slot for her "last message" indicates that this broadcast was fully excised, for some reason, *only after the Navy was apprised of matters*, at 2:05am, on July 3.

Form 2614 A—Revised Oct. 1933
TREASURY DEPARTMENT
U. S. COAST GUARD

RADIO LOG .... **ITASCA** .... Date .... **2 JULY, 1937**

ENTRIES	TIME
W L GALTEN, RM3C ON - CGR-32-1	
KHAQQ DE NRUI GA 500 WID DASHES ON 500 K / UNANSWD	0805-8
KHAQQ FM ITASCA DID U GET TT XMISION ON 7.5 MEGS GA ON 500 KCS SO TT WE MAY BE ABLE TO TAKE A BEARING ON U IMPOSSIBLE TO TAKE A BEARING ON 3105 - PLS ACKNWLDGE THIS XMISION WID A3 ON 3105 GA / UNANSWD	11
KHAQQ DE NRUI RPTING ABVE INFO ON 7500 / NO ANSWR	12-14
KHAQQ FM ITASCA DO U HR MY SIGS ON 7500 KCS OR 3105 KCS PLS ACKNWLDGE WID RECEIPT ON 3105 KCS WID A3 GA / UNANSWD	15
KHAQQ DE NRUI REPTED ABVE DPE ON 7500 GA 3105 A3 AR / UNANSWD	16-7
KHAQQ FM ITASCA WL U PLS Y OUR SIGS ON 7500 OR 3105 GA WID 3105 A3	18
KHAQQ UNANSWD	19
KHAQQ DE NRUI GA 3105 A3 KCS WID REPORT OUR SIGS	20-3
KHAQQ DE NRUI GA 3105 KCS WID A3 ES XMIT POSN REPT ES QSA ON OUR SIGS	24-6
ITASCA TO EARHART WE XMITING CONSTANTLY ON 7.5 MEGS DO U HR US KINDLY CFM RECEIPT ON 3105 WE ARE STANDING BY, A3/3105	27
KHAQQ DE NRUI ANS 3105 A3 K / UNANSWD	28-9
KHAQQ DE NRUI ANS 3105 KCS WITH REPORT ES POSN, XXES 7500 / UNANSWD	30-1
EARHART FM ITASCA WL U PLS CUM IN AND ANS ON 3105 WE ARE XMITING CONSTANTLY ON 7500 KCS WE DO NOT HR U ON 3105 PLS AND 3105 GA / UNANSWD	33
KHAQQ DE NRUI ANS 3105 KCS WID A3 HW OUR SIG QSA? GA / UNANSWD	34-41
LSNIN 3105 / NIL - CRM TUNIN UP T16 FER XMISION TO NMC	42
NMC V NRUI P AR, 12600 / UNANSWD	43
KHAQQ TO ITASCA WE ARE ON THE LINE 157 337 WE WL REPT MSG WE WL REPT N ES S THIS ON 6210 KCS WAIT, 3105/A3 S5 (?/KHAQQ XMISION WE ARE RUNNING ON XX LINE	43
LSNIN 6210 KCS / KHAQQ DE NRUI HRD U OK ON 3105 KCS RX, 7500	44-6
KHAQQ DE NRUI PLS STAY ON 3105 KCS DO NOT HR U ON 6210 MAINTAIN QSO ON 3105, 7500 / UNANSWD	47
NIL ON 3105 OR XX 6210 FM KHAQQ / KHAQQ DE NRUI ANS 3105 KCS	48
KHAQQ DE NRUI ANS 3105 KCS A3 / UNANSWD	49-53
KHAQQ DE NRUI UR SIGS OK ON 3105 GO AHEAD WITH POSN ON 3105 OR 500 KCS / UNANSWD	54-0907
LSNIN 3105 AND 6210/500 KCS / NIL	08
KHAQQ DE NRUI ANS 3105 OR 500 UR SIGS OK ON 3105 GA WID POSN, 7500	09-13
LSNIN 6210 ES 3105 - NIL / 500 NIL	14
LSNIN 3105, 6210, 500 AND 500 D/F - NIL	15-33
NRUI2 V NRUI - PER TO, REF HI FREQUENCY D/F, 7500	34
KHAQQ DE NRUI GA ON 3105 KCS, 7500 / UNANSWD	35
NIL FROM KHAQQ / 3105, 6210, 500 OR 500 D/F	36-41
KHAQQ DE NRUI: XMITING DPE ON ANSWING FREQS / WE CAN HEAR U FINE ON 3105 PLS GA ON 3105, 7500 / UNANSWD	42-6
NIL FROM KHAQQ	49
NIL FROM KHAQQ / 3105, 6210, 500 OR 500 D/F	52-9
NRUI2 V NRUI: - PER TO / GET THE RDO COMPASS WRKING NOW, 7500	1002-2
KHAQQ DE NRUI WE HEARD YOU ON 3105 KCS, ETC - 8500	03-5
NPM NPU V NRUI V-S ZME / NPM V NRUI INT ZCT KHAQQ INT ZCY KHAQQ AR	08-10
NRUI V NPM ZZA INT FREQ AR / 3105 / R ZOZ AR / DID U HERE KHAQQ AT ALL AR / ZZA /R THAT T TO NPU / R (NPM/NRUI - 12600/13380) - PER TO	12
NRUI V NPM AR / NPM V NRUI RK / BMBR3 B NR3 / R AS	14
TO ENGAGED NPM: XMITING P MSGS, 12600	15-26
DC ENGAGED NPM: RECVING P , 12600 / LSNIN 3105 CONSTANTLY, NIL / BC	27-30
BC OFF TO DC//	35
WGEN WGEN DE NRUI NRUI QRU IMI	38
WGEN WGEN DE NRUI QRU 600 K	39

(OVER)

Original Radio Log – USCGC *Itasca*

Original Log page from radio position number 2, the special Earhart position, 8:05am to 10:39am, July 2, 1937 (in a private collection until 1975).

EARHART'S LAST MESSAGE(S)

Referring to the radio log page from *Itasca*'s special Earhart radio position, the message logged during minute 43 may be something of a surprise to the reader. But the first part of the message is simple enough: "KHAQQ TO ITASCA WE ARE ON THE LINE 157 337 ~~WE~~ WL REPT MSG WE WL REPT THIS ON 6210 KCS WAIT, 3105/A3 S5."

The translation is simple, as well. KHAQQ was Earhart's call sign (although KHABQ had been her call sign two years before, in 1935, for her old Lockheed *Vega*): "Earhart to *Itasca*. We are on the reciprocal line of position 157-337. Will repeat message. We will repeat this on 6210 kcs [kHz]. Wait." Following the message text is a notation of the frequency of transmission (3105kHz), the type of emission (voice modulation, A3) and the signal strength (the highest, S5), as noted by Bill Galten, the operator on watch.

The only problem with this message was the order to "wait," at its end. For Bill Galten, it may have posed a momentary dilemma, as an order to wait (until the other operator gets back to you) is *never* to be violated. In fact, one of the radiomen aboard *Itasca*, Gerald Thompson, had been brought up on charges, a few days before, on this very issue. On the other hand, Earhart was not a real operator, and, based upon her expressed intention to repeat the message on 6210kHz, Galten may well have felt compelled to try to confirm the message by monitoring 6210kHz, her daytime frequency. His next entry, "LSNIN 6210 KCS [Listening on 6210kHz]" seems to confirm that he *did* shift to 6210kHz. But it will be noted that someone, certainly from the *Itasca*'s crew—probably Bellarts—has underscored the word "wait," undoubtedly for the reasons cited above. Technically, the order to wait was valid, but, practically, it made no sense, and the next entry, a plea to Earhart to "stay on 3105 kcs," was useless, as it was sent on 7500kHz, using wireless telegraphy (indicated by the form, "KHAQQ DE NRUI"), which Earhart could not copy. Nor did Earhart, so far as is known, read any "Q" signals, such as, "MAINTAIN QSO ON 3105," which Galten included at the end of his 8:47am plea. It is clear that the radiomen of *Itasca* expected to be communicating, not only with someone who could "read Morse," but with a genuine, experienced commercial wireless telegraph operator, totally familiar with "Q" signals—such as Harry Balfour.

There is yet another matter regarding the 8:43am message entry. The reader will note that, prior to the Earhart entry at 8:43am, there are two entries pertaining to an attempt to communicate with NMC (Coast Guard San Francisco Division). These entries were logged originally at 8:44am and 8:45am, but with no contact having been made. The two entry times have been struck through, and the times changed to 8:42am and 8:43am, respectively, a fairly tight fit in this log.

But 8:44am and 8:45am were inappropriate times to be firing up the T-16 transmitter, as Earhart's schedule, to which she adhered, for the most part (sometimes coming in a bit early), called for communication at quarter past, and quarter to, the hour. It seems likely that Earhart contacted *Itasca* at 8:43am, as ultimately logged, but that Bill Galten, for whatever reasons, failed to log her message. At 8:44, after she had signed off, Bellarts may have been ordered to let San Francisco know that she had called in. The attempt to contact NMC was unsuccessful, and, by the time that Galten had gotten back to his log (after a plea to Earhart to stay on 3105kHz), he logged the NMC contact attempt on the next line *before* logging Earhart's message. By juggling the times, he managed to log Earhart at the correct time, 8:43am, without completely fouling the log, but the attempt to contact NMC wound up being logged as occurring *before* the receipt of Earhart's message. (The transcript would "correct" the error by logging Earhart in at *8:46am*.)

TH REPORT ES POSN, XXXX 7500 / UNANSWD			30-1
M IN AND ANS ON 3105 WE ARE XMITING CONS-			
HR U ON 3105 PLS AND 3105 GA / UNANSWD			33 D
D A3 HW OUR SIG QSA? GA / UNANSWD			34-41
UP T16 FER XMISION TO NMC			XX 42
SWD ,A		B C	XX-43
LINE E157 337 XX WL REPT MSG WE WL REPT N ES S'			
A3 S5F(?/KHAQQ XMISION WE ARE RUNNING ON XX LINE			
I HRD U OK ON 3105 KCS RX, 7500			44-6 43
5 KCS GDO NOT HR U ON 6210 MAINTAIN QSO			G
			47
QQ / KHAQQ DE NRUI ANS 3105 KCS			48
/ UNANSWD			49-53
105 GO AHEAD WITH POSN ON 3105 OR 500			
			54-0907

So much for the first part of Earhart's last message. The second part is mind-boggling.

The second part of the message is difficult to present because it cannot simply be typed, character for character or word for word. The reader will note, on the log sheet, that, on the second *line* of the 8:43am entry, the second *part* reads, "(? / KHAQQ XMISION WE ARE RUNNING ON 43 LINE," with the phrase "N ES S," i.e., *North and South*, crammed into the space above "43 LINE."

The question mark, at the beginning of this second part of Earhart's message, indicates that essential data for this part is unknown; that is, frequency, signal strength, perhaps even the time. "XMISION" is *transmission*, of course, and "43" is the time of receipt of Earhart's original first message, at 8:43am. This entry has been struck over to eliminate confusion with the second part, in the last message. The original time, 8:43am ("43"), was then added a half-space below "LINE," the last word on the second line.

Since the author has presented this message in two parts, it should be noted that this is not the first time that it has been considered as a two-part message, or that the question of time has been broached. On page 45 of the *Itasca*'s long report, "Radio Transcripts, Earhart Flight," the message is presented as two parts, copied from 8:44am to *8:46am* (the time which would have been logged, as being after the actual NMC contact attempt, had Galten not altered the NMC times). This was because the report was assembled from a *transcript* of the original log, in which an attempt was made to "reconstruct" the times.

Because there was no indentation of lines in the transcript, the first part of her message, *as printed in the report*, includes the garbled phrase "wait listening on 6210 kcs," which is Earhart's "wait" and *Itasca*'s "lsnin 6210 kcs," run together. Yet, the report notes that, "Other persons in radio room heard this transmission the same," which is *simply not true*, as can be seen on the original log sheet. The *north and south* comment is separate from the first part, and appears, in the report, *after* the "other persons" note, as, "We are running north and south," but *without the phrase*, "on line," which does appear in the log transcript, the source for the report.

The point of all this is not that *Itasca* was simply "gun-decking" (fabricating) her logs, but that there was *unsureness* aboard *Itasca*, regarding the *north and south* comment.

A blow-up of the original log sheet has been provided in order to demonstrate another unusual feature of the second part of the last message. Close examination of this blow-up reveals that the separate phrases, "WE ARE RUNNING ON XX LINE," and "N ES S," are sitting above the print line of the earlier entries, as indicated by the underscoring of these phrases and the horizontal reference lines, E and F. The reference lines D and G show that the surrounding entries, before and after the 8:43am entry, are relatively even across the page. Further, the vertical reference lines, A, B and C, reveal that the entire second part of the message, including the technical data entry at its beginning, is offset, in the horizontal plane, by half a character width. The clearest example of the offsets is the strike-over of the original time, minute "43," between "ON" and "LINE."

Anyone who is familiar with typewriters will recognize that these line offsets are the result of the sheet having been removed and *re-inserted* into the typewriter, certainly no earlier than the 9:09am entry, and certainly no later than the drafting of the message which *Itasca* sent out at 8:15pm. In adding this second part of Earhart's final message, an attempt was apparently made to "float" the platen in the vertical plane, but, as typing proceeded, past the technical data, the platen rotated until it latched into a detent position, with the resulting slight vertical offset.

The second part of the last message reached the officers at a later time, certainly between 2:02pm and 8:15pm on the day of her loss. This is clearly indicated by the references to a last Earhart contact *at 8:55am*, included, along with the *north and south* comment, in the two later messages sent from *Itasca*, at 8:15pm (on July 2) and 2:05am (on July 3).

As to why the second part of Earhart's final message was nearly lost, it probably turns on the apparent fact that Earhart had quite a lot to say (possibly without knowing whether *Itasca* could hear her), from 8:55am until she ditched, at about 9:05am. Because radio logs are intended to contain only "pertinent data," that is, technically useful data, all of this final transmission was left out of, or expunged from, the official records, and the 8:55am time slot was left *essentially* empty.

It will be noted, however, that this time slot extends from 8:54am (the call to Earhart), through *8:55am* and on to *9:07am*, making it, at thirteen minutes, the longest "empty" block of logged time on the sheet. On the log for the *Itasca*'s main position, O'Hare noted Galten as logging all "pertinent data" at *8:56am* (followed by his own *delayed entry* of Earhart's 8:00am transmission). He then logged a long block of contact-attempt activity, which he noted as having been terminated, at *9:08am*. Clearly, as witnesses have noted, and as the logs and the outgoing message summaries confirm, there was a long broadcast series from Earhart, between 8:55/:56am and 9:07/:08am, with a clearly disturbing content and little "pertinent data."

It seems likely that the *north and south* comment was regarded, at 8:43am, merely as a comment on the "search tactic" which *Itasca* wrongly assumed that Earhart was employing, that is, flying "north and south" along the line of position, hoping to locate Howland Island (wrongly assumed to be her reference point). When it was repeated at 8:55am, probably along with a repeat of the Line of Position data, it may have registered, in the minds of the officers, as a pointless comment for Earhart to have repeated, since, after 8:55am, she was so close to ditching. Later on, in the *post mortem* which always follows events of this nature, the strangely repeated comment may have been discussed, and, on the chance that it might be pertinent, it was crammed into the log, long afterward.

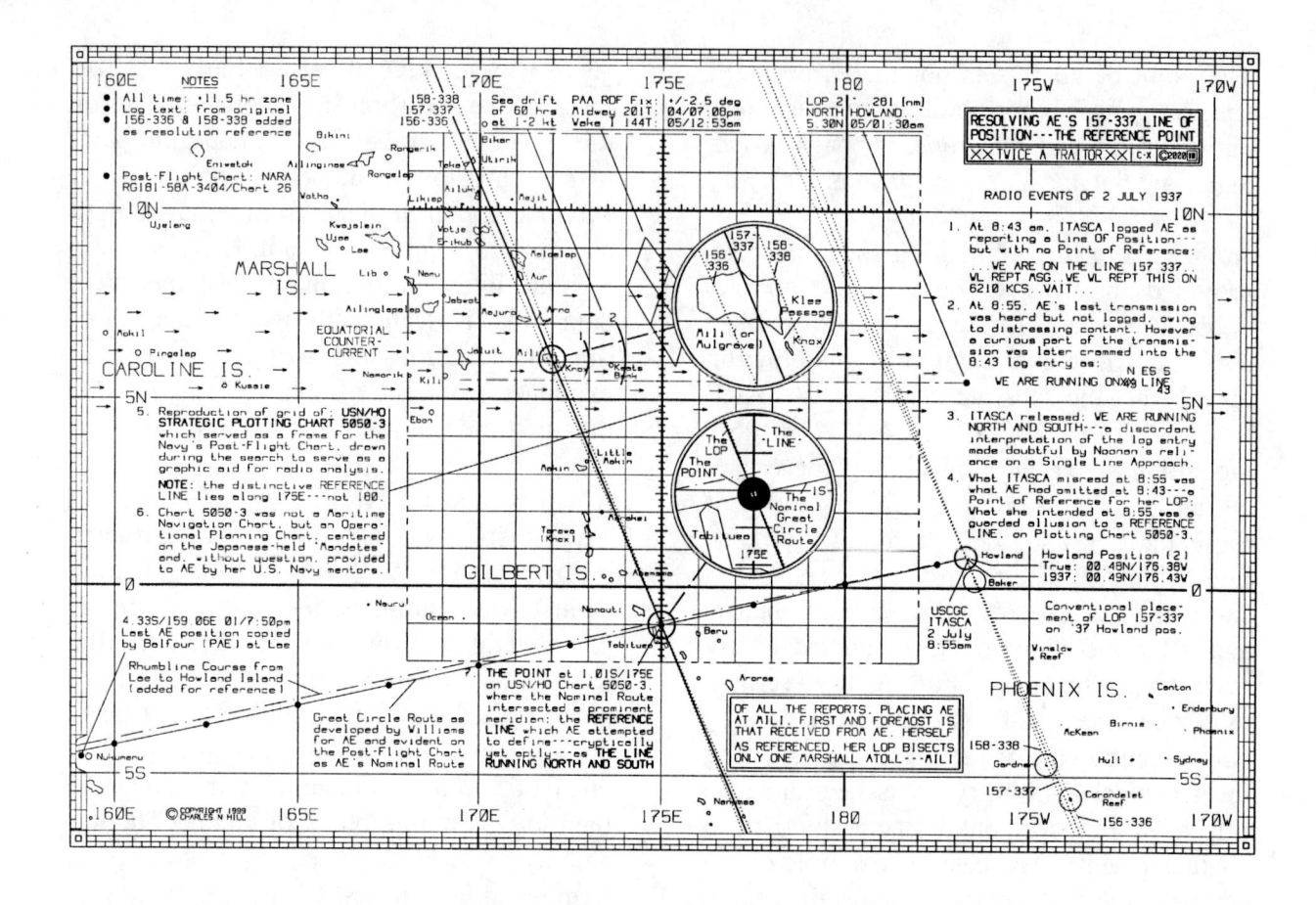

It was quite pertinent. In the first two of the four outgoing messages cited above (that is, those which were sent out at 10:15am and 2:02pm respectively), *Itasca* made a specific reference to the fact that the 157-337 Line of Position was broadcast by Earhart *without a reference point*. But, of course, she had not neglected the reference point for her Line of Position. The *north and south* comment, which had very nearly been discarded, was the *key*, albeit a cryptic key, to the location of that reference point, and that point was *not* Howland Island, as *Itasca* had assumed.

What Earhart was attempting to inform the *Itasca*, so desperately, but so discreetly (and, certainly, *too* discreetly), was not that she was running *north and south* on the Line of Position (at a time when she was preparing to *ditch*, for God's sake), but that her Line of Position, and its vital reference point, were functionally tied, *on a specific Navy chart*, to—"*the line* running north and south."

The chart which was used aboard *Itasca*, for the purpose of plotting the long approach to Howland Island, and later, for the evaluation of her possible locations after ditching, was *Strategic Plotting Chart 5050-3*, published by the U.S. Navy Hydrographic Office. This was not a marine navigation chart at all, as it was uncluttered with the information and data which are required for navigation and which are invariably included on navigation charts. Strategic plotting charts were simple, sea charts depicting land masses, islands and surface reefs, with a dense grid of latitudinal and longitudinal lines, whose density would actually clutter a navigation chart. Using this chart (similar to the author's, above), *Itasca* had originally expressed a belief that the reference point for Earhart's 157-337 Line of Position was probably Howland Island and that she had gone down near this line some 40 miles to the north-northwest. Initially, this was a very logical assumption, which deserved to be examined closely and then discarded (as it actually was), when no debris was found during the first search of the sea to the north-northwest of Howland. However, as is shown on the chart, at left, the 157-337 Line of Position passes near Gardner Island and Carondelet Reef, far to the south-southeast (and Baker, which was in contact with *Itasca*). This led to the false theory, current even today, that Earhart might have ditched at one of the Phoenix Islands. The U.S. Navy mounted its first search in these islands and a private group has offered useless "expeditions" to Gardner (now Nikumaroro) during the past decade.

However, the chart, at left, includes 158-338 and 156-336 Lines of Position, as well—as a test of resolution for the 157-337 line that Earhart provided. It can be seen that the 158-338 line accurately crosses Gardner Island and that the 156-336 line accurately crosses Carondelet Reef. Had she ditched near either one of these locations, she would have used the proper line for the location, not the 157-337. However, this does indicate that searchers, in 1937, had assumed that she may have specified the 157-337, not merely as a *Line* of Position, on which she could be *anywhere*, but as a *pointer*, radiating from a reference point to a *specific location*. This was a proper assumption, as will be shown.

The chart, at left, also shows the two routes from Lae, New Guinea to Howland Island. The dashed line shows a single Rhumbline Course over the route (for reference), while the solid line shows the *official* route, which was planned by Clarence Williams. This is a Great Circle Route, made up of a series of Rhumblines, connected end to end. Each of the "dots" represents the connection point of a Rhumbline pair. Williams specified each of these positions as part of a dead-reckoned proposed route, that is, a *nominal* route, with no winds specified, for obvious reasons.

Two reasonable assumptions were made, in 1937, regarding Earhart's Line of Position:

First, and following the customary practices of navigation, it was assumed that Earhart had referenced her Line of Position from a specific point *on her dead-reckoned nominal route* (such as her destination, at Howland Island). Second, it was assumed that she had intended that her reported Line of Position serve as a pointer to a *specific* location.

With this in mind, Earhart's location can be determined. Strategic Plotting Chart 5050-3, which covers the former Japanese Mandates, is curious in one respect: the most prominent longitudinal line on the eastern half of the chart, is not, as one might expect, Longitude 180 East/West, the International Date Line. Instead, it is a bold, heavy line, indexed with latitudinal subdivisions, lying just to the east of the Japanese Marshalls, at Longitude *175 East*. A small portion of the grid of Strategic Plotting Chart 5050-3 has been reproduced on the chart at left, with absolute accuracy as to position and line density. Her reference point for the 157-337 Line of Position was the point of intersection of the original route, *on the line running north and south*, that is, on the only *highly prominent* longitudinal line on the eastern half of the plotting chart.

The 157-337 Line of Position, referenced to that point, Latitude 1.01S, Longitude 175E, *grazes* a number of islands in the Marshalls group, but passes *directly* through only one: *Mili Atoll*. Earhart used the reciprocal line of 157-337, rather than the direct line of 337, to promote a *general* search, rather than a concentrated search northwest of Howland. Despite the failure of her overly cryptic plea, Earhart provided the *very first* of the many reports, from 1937 into the 1980's, placing her at Mili Atoll, in the Japanese Marshalls.

As to her fate, the author is informed that "*she who was and always will be*," born July 24, 1897, did cease to *be*, on May 20, 1999, at the remarkable age of 101 years, 9 months and 27 days, her ashes being scattered, at her request, over the Palau and Marshall Islands.

4. THE SAILINGS: Analytical Navigation Mathematics—the Equations and the Data

INTRODUCTION

Sailings is a nautical term for navigational directions, which can be expressed as textual instructions or as mathematical equations. *Sailings* provided a means of navigation in the times prior to Mercator's work and the printing of large charts, which ultimately provided the now familiar graphic system. Originally, the term *sailings* applied only to the detailed instructions as to distance and course for dead-reckoning voyages between specific ports. These written directions also indicated the location of shoals, reefs and landmarks, for pilotage in specific sea areas.

Even during the Age of Discovery, as charts did become available, the written *sailings* of the trade route explorers could be collected, compiled and then printed in book form, for simpler dissemination, easier correction and greater protection (during the competitive development of international trade routes) than was possible with large charts. Today, *sailings* are used in written or textual form for every official *Notice to Mariners*, issued to warn of hazards along the sea lanes.

As a knowledge of both plane and spherical trigonometry became more common among navigators, navigational equations also came to be considered as *sailings*, but it was not until the development of logarithmic slide rules and the advent of fast, long range air transport along Great Circle Routes that the direct application of such equations became both practical and quite necessary. Today, as navigation rapidly becomes a lost art, these most abstract of all *sailings* have found their ultimate expression as the soul of electronic navigation systems, or, as in the case of the present work, as the means of preparing the series of analytical charts required to define the *Hawaii Clipper* hi-jacking.

However, no reader should feel intimidated by the mathematics in this work. Even with the decline of navigation as a profession, the daily use of graphic methods of navigation, including a limited reliance upon abstract mathematics, remains both a practical and familiar reality for most highway travelers:

For example, for a trip, in 1996, to the new home of my long-time friend, Mrs. Sherry Holdsworth, of New Philadelphia, Ohio, a road atlas *chart* provided a graphic means of navigating the interstate system between Cincinnati and New Philadelphia. However, locating Sherry's residence in that town also required detailed directions, describing the direct route from the I-77 exit to her home.

More practical than even a city street map, Sherry's verbal *sailings*, provided to me by telephone, provided landmark references for *pilotage*, distance estimates along the route for *dead-reckoning* the positions of cross streets, and additional details required for an evening ETA (Estimated Time of Arrival), when street signs and house numbers would not be clearly visible from my car. She even provided an *aid-to-navigation*, in the form of a lighted porch lamp, but it was her concise directions, or *sailings*, which made the trip both simple and direct.

During the trip, my car even experienced the same force as that experienced in air travel: the wind vector. On the plains to the south of Columbus, Ohio, a strong west wind (a wind blowing *from* the west) caused the car to drift toward the right, requiring a slight leftward pressure on the steering wheel. This leftward heading compensated for the wind and kept the car on the course of the road, and, east of Columbus, the westerly tailwind helped to push the car along, reducing the fuel required to reach I-77 at Cambridge.

In a car, of course, winds demand no special navigation skills. However, planning even such a simple trip involved the limited use of basic navigational mathematics, if only to calculate the Estimated Time of Arrival (by calculating the time [T] required for the trip from estimates of distance [M] and speed [K], as in T = M/K.) Twice the value of the one-way distance, divided by the known fuel mileage of the car, provided an estimate of the total amount of gasoline required for the round trip, which, when multiplied by its expected cost, per gallon, also indicated the minimum cost of the trip. These are the most common of the abstract sailings utilized in highway travel, but it is not uncommon for travelers to analyze a trip, once it has been completed, in order to confirm, by simple mathematics, their car's actual mileage, or to compare the true effectiveness of one route with that of another. And, while the abstract mathematical *sailings* used by most highway travelers are far less sophisticated than those used in this analysis, the basic principle of the application of abstract mathematics is essentially the same in both cases.

PURPOSE: Although many of the sailings contained in this Section involve equations of plane and spherical trigonometry and are of a far higher order than the simple and familiar equation noted above (although it, too, played a part in this analysis), it was never my purpose to provide a formal course in dead-reckoning navigation for the reader. Rather, this section of the general Appendix was included to provide a means by which readers having skill in the art, and having access to the same tools that the author used, might confirm, for themselves, the accuracy of the *Hawaii Clipper* flight analysis.

SCOPE: The scope of this Appendix extends only to the limits required to provide support for the specific analysis under consideration and may well prove insufficient as a means of providing support for analyses of other examples dead-reckoned navigation.

GLOSSARY OF NAVIGATION TERMS

PILOTAGE: visual navigation along known routes by reference to natural and man-made landmarks and to fixed aids-to-navigation.

CELESTIAL NAVIGATION: determination of one's position or placement along a line of position, by measured observation of the Sun, the Moon, the planets or the fixed stars.

DEAD-RECKONING: a practical, although theoretical, flight plan, which provides a true heading based solely upon an observation or determination of certain specific variables, including: geographic positions of departure and arrival; the working altitude and true air speed of the aircraft; the wind speed and direction at cruise altitude; the temperature and barometric pressure at the cruise altitude and at sea level; and the magnetic compass variations along the route (not required for this analysis).

TRUE BEARING ANGLE: a directional angle expressed with reference to the axis of rotation of the globe and defined by the intersection of any Meridian (of longitude) passing through a fixed point on a Mercator chart and a line segment extending in any direction from that point and measured, in decimal degrees, from the Meridian segment to the north of the fixed point, clockwise to the line (less than 360 degrees True).

LOXODROME: a line on the surface of the globe that intersects all lines of latitude or longitude at the same True Bearing Angle. On the Mercator-type navigation charts, the mathematical distortion of latitude converts generally curved loxodrome into straight lines, which express a constant True Bearing Angle between any two points along the line. Although seldom the shortest distance between points (except in a few well-defined special cases), a loxodrome segment is the simplest course to plot on a Mercator chart, owing to its constant True Bearing Angle.

GREAT CIRCLE: any circle girdling the (theoretically spherical) globe at its surface, the locus of whose center is the center of the globe. A Great Circle passing through fixed points on the surface of the globe constitutes the shortest distance between those points, but the True Bearing Angle of any Great Circle, other than that of the Equator or any Meridian, is infinitely variable and therefore unsuitable as a practical course or heading.

RHUMBLINE: a segment of a Loxodrome between two points on a Mercator chart, being of such a relatively short distance, for its latitude, that it is not appreciably longer than a Great Circle segment between the two points. It is, therefore, practical as a fixed course or heading for transport between the two points. For great distances or for courses in the higher latitudes, where a Great Circle generally represent a significant reduction in distance over a Rhumbline, a Great Circle may be approximated by a contiguous series of shorter Rhumblines (having distinct True Bearing angles), which are plotted point-to-point along the Great Circle course.

MERCATOR PROJECTION: The Mercator Projection provides a means of expressing the near-spherical globe in two dimensions; that is, as a chart, printed on paper. Its value to navigation lies in its distortion of latitude, which permits all Rhumblines (with constant True Bearing Angles, or compass headings) to be drawn as *straight lines* between any two points on the chart. Not only is this a convenience for plotting courses, but a boon to sailing them, as well, since, theoretically, any point on the globe can be reached from any other point by traveling along a *single* True Bearing Angle (or corrected compass course). In practice, longer voyages are best plotted as Great Circle routes, then broken into a series of Rhumblines along the routes. The Mercator Projection was discovered by Gerardus Mercator (1512-1594), a Flemish cartographer, who introduced the projection in his first world map, published in 1569. The use of the word "projection" derives from the geometric means of defining such a world-view, which, in the case of Mercator projections, is classified as "cylindrical:"

One may imagine a globe of the earth, made from clear glass or plastic, with land features and lines of latitude and longitude drawn on its surface. The globe is mounted so that its axis of rotation (pole-to-pole) is vertically oriented, with a pinpoint light installed at the globe's center. One may imagine turning off the ambient lighting and wrapping one large sheet of photographic paper around the globe, so that the axis of the globe matches the axis of the "cylinder" of paper, with the sensitized side of the paper facing the globe. If the light is energized, for just a moment, the paper can be removed and "developed," revealing the image of the globe's features, which have been "projected" onto the paper.

The image would be a Mercator "chart," but it would take a very long cylinder of paper to provide coverage of the bulk of the globe, especially in the higher latitudes, and, as the poles lie on the globe's axis, there can never be a projection of the poles on the cylinder.

The *mathematical* expression for Mercator's projection is given by the equation below, where C equals either one limit of the chart, or any desired latitude, expressed in decimal degrees; where A equals the other limit of the chart, or a reference latitude, expressed in decimal degrees; and where W is the chart scale in units such as "one inch per degree." Y equals the distance, in such units, between latitudes C and A, where C and A are each *less than* ninety degrees. For a scale of one inch per degree, a "pole-to-pole" world chart (+/- 89 degrees) would be 180 inches wide and over *543* inches high! At +/- *80* degrees, it would be much less, only 279 inches high, becoming more linear near the Equator:

(19) Y = (LNTAN(45+C/2)-LNTAN(45 +A/2))* W*180/PI

MISCELLANEOUS DATA

POSITIONS: The following positional data provide the frame of reference for this work, and are expressed in Degrees, Minutes and Seconds (DMS) of arc, to the N(orth) of the Equator, as Latitude, or, to the E(ast) of the Prime Meridian, as Longitude.

There are 60 seconds in a minute of arc and 60 minutes in a degree, and a minute of arc along any great circle may be expressed as a nautical mile. As there are 360 degrees in a circle, Latitude is limited to a maximum of 90 degrees, north or south of the Equator, and Longitude is limited to a maximum of 180 degrees, east or west of the Prime Meridian (at Greenwich). In converting the DMS values to Decimal Degrees (DEC) or to Radians (RAD) for computation, northern latitudes and eastern longitudes are both expressed as positive decimal degree values, as are displacements (or vector components) in those directions. The values are negative for southern latitudes, western longitudes and for their respective displacements. As is noted elsewhere, angles must be converted to radians (RAD) for calculation in BASIC.

Proper names and data for all of the known geographic positions reflect common usage of the late 1930's, while data for the flight positions have been derived from reports (in brackets), or by a mathematical analysis of reported flight and weather data.

POSITION		LATITUDE	LONGITUDE
Chart Origin		00:00:00:N	119:00:00:E
Chart Limits		22:00:00:N	153:00:00:E
Agana, Guam		13:28:30:N	144:44:50:E
Apra Hbr, Guam		13:27:00:N	144:39:00:E
Canacao Bay, P.I.		14:29:30:N	120:54:50:E
Cavite, P.I.		14:29:00:N	120:54:40:E
Dublon I., Truk		07:22:45:N	151:52:45:E
Dublon S.P. Base		07:21:50:N	151:52:45:E
Garapan, Saipan		15:12:45:N	145:43:25:E
Hosp. Slab, Truk		07:22:32:N	151:53:55:E
Hong Kong		22:15:00:N	114:10:00:E
Infanta, P.I.		14:45:00:N	121:39:00:E
Ise Bay, Japan		35:00:00:N	136:50:00:E
KNBG Sumay		13:26:15:N	144:39:05:E
Koror, Palau		07:20:15:N	134:28:30:E
KZBQ Makati RDF		14:33:30:N	121:01:40:E
KZDY Panay		13:58:00:N	124:20:00:E
Lahuy I., P.I.		13:56:00:N	123:50:00:E
Landing: Apra		13:27:00:N	144:39:00:E
Landing: Cavite		14:29:28:N	120:55:38:E
Landing: Eten		07:20:30:N	151:52:30:E
Landing: Malakal		07:19:00:N	134:28:00:E
Landing: Manila		14:35:00:N	120:57:00:E
Landing: Tanapag		15:13:45:N	145:43:45:E
Landing: Ulithi		10:04:00:N	139:44:00:E
Laoang, P.I.		12:35:00:N	125:00:00:E
Manila Bay, P.I.		14:30:00:N	120:45:00:E
Manila, P.I.		14:36:00:N	120:58:00:E
Nagoya, Japan		35:10:00:N	136:55:00:E
Orote Pt., Guam		13:26:50:N	144:37:10:E
PAA, Cavite		14:29:06:N	120:54:05:E
PAA, Sumay		13:26:15:N	144:39:05:E
Pagan, Marianas		18:03:00:N	145:46:00:E
Parece Vela		20:24:00:N	136:02:00:E
San Fernando, P.I.		16:37:00:N	120:19:00:E
Sangley Pt., P.I.		14:30:00:N	120:55:00:E
Sumay, Guam		13:26:00:N	144:39:00:E
Takao, Formosa		22:36:00:N	120:17:00:E
Tanapag SP Base		15:13:30:N	145:44:25:E
Tanapag, Saipan		15:14:30:N	145:45:40:E
Wake I.		19:18:00:N	166:37:00:E
H6K: HC Intercept		14:17:00:N	128:07:00:E
H6K: Lahuy Turn		13:52:00:N	123:50:00:E
H6K:	0404Z	12:40:00:N	129:36:00:E
Meigs	0349Z	13:01:00:N	129:12:00:E
Meigs	0404Z	13:01:00:N	129:14:00:E
Meigs	[0411Z]	13:01:00:N	129:15:00:E
Meigs	0740Z	13:02:00:N	129:47:00:E
Meigs	[Oil Slick]	12:11:00:N	130:33:00:E
Mid-Course Point		13:58:00:N	134:30:00:E
Panay: Check Pt.		14:29:00:N	124:20:00:E
Pos. A	[0200Z]	12:20:00:N	134:26:00:E
Pos. A1	0130Z	12:30:00:N	134:25:00:E
Pos. B	[0230Z]	12:00:00:N	133:36:00:E
Pos. B0	0230Z	11:36:00:N	135:20:00:E
Pos. B1	0230Z	12:00:00:N	135:36:00:E
Pos. C	[0300Z]	12:05:00:N	133:28:00:E
Pos. C0	0300Z	11:06:00:N	136:26:00:E
Pos. C1	0300Z	12:05:00:N	132:28:00:E
Pos. D	[0330Z]	12:15:00:N	131:37:00:E
Pos. D0	0330Z	10:45:00:N	137:36:00:E
Pos. D1	0300Z	10:46:00:N	137:29:00:E
Pos. E	[0400Z]	12:27:00:N	130:40:00:E
Pos. E0	0400Z	10:23:00:N	138:45:00:E
Pos. E1	0330Z	12:27:00:N	130:41:00:E
Pos. E2	0330Z	10:25:00:N	138:38:00:E
Pos. F	0411Z	12:31:00:N	130:20:00:E
Pos. F0	0411Z	10:15:00:N	139:10:00:E
Pos. F1	0344Z	10:15:00:N	139:10:00:E
Pos. F2	Oil Drop	12:33:00:N	130:14:00:E
Pos. P0	2200Z	13:38:00:N	140:52:00:E
Pos. P1	2030Z	13:29:00:N	143:59:00:F
Pos. P2	2100Z	13:32:00:N	143:06:00:E
Pos. P3	2130Z	13:35:00:N	142:12:00:E
Pos. P4	2200Z	13:37:00:N	141:18:00:E
Pos. P5	2230Z	13:40:00:N	140:24:00:E
Pos. P6	2300Z	13:43:00:N	139:30:00:E
Pos. P7	2330Z	13:46:00:N	138:36:00:E
Pos. P8	0000Z	13:48:00:N	137:42:00:E
Pos. P9	0030Z	13:26:00:N	136:53:00:E
Pos. P10	0100Z	13:04:00:N	136:04:00:E
Pos. P11	0130Z	12:42:00:N	135:15:00:E
Pos. U0	0426Z	10:04:00:N	139:44:00:E
Term A-180T	0200Z	07:20:00:N	134:26:00:E
Term A-B1	0230Z	07:18:52:N	151:52:01:E
Term C-A	0300Z	15:14:07:N	145:43:39:E
Term D-E	0400Z	14:29:29:N	120:55:36:E

CONTEMPORARY FLIGHT DATA:

```
A. Distance, Guam-Manila:       1589sm; 1380nm
B. Estimated Flight Time:           12hr, 30mn
C. Max. Range, Trip 229:    2600sm; 2257.74nm
D. Endurance, 7800ft Density Alt: 17hr, 30mn
E. Altitude 290400Z:                     9100ft
F. Temperature at 9100ft 290400Z:           13C
G. Wind Speed 290400Z:                     19kt
H. Wind Dir. 290400Z:       247 deg True: 247T
I. Ground Speed 290400Z:                  112kt
J. Desired Track 290400Z: 282 deg True: 282T
K. RDF Bearing from Makati 290345Z:        101T
L. Wind, Guam at 8000ft 281830Z:     WSW, 19kt
M. Wind, Cavite at 8000ft 281830Z: WSW, 20kt
N. Sea Level Temp. Guam 281939Z:            80F
O. Sea Level Temp. Cavite 281939Z:          81F
P. S.L. Pressure, Guam 281939Z:    29.83 in Hg
Q. S.L. Pressure, Cavite 281939Z: 29.76 in Hg
R. Ceiling Guam 281939Z    2000ft 5/10 Clouds
S. Ceiling Cavite 281939Z  2500ft 1/10 Clouds
```

UNIT ABBREVIATIONS:

```
A. kt: knot(s), nautical mile(s) per hour
B. nm: nautical mile(s)
C. sm: statute mile(s)
D. hr: hour(s)
E. mn: minute(s)
F. sc: second(s)
G. #T: decimal degrees True, as in 247T
H. in: inch(es)
I. ft: foot(feet)
J. am: ante meridian (before noon)
K. pm: post meridian (after noon)
L. #F: decimal degrees Fahrenheit
M. #C: decimal degrees Centigrade/Celsius
```

FUNCTION DESIGNATORS:

A through T are fixed-function designators.
U through Z are variable-function designators.

```
A = Latitude, Pos 1, in decimal degrees
B = Longitude, Pos 1, in decimal degrees
C = Latitude, Pos 2, given or calculated
D = Longitude, Pos 2, given or calculated
E = Latitude, Pos 3, calc. or intermediate
F = Longitude, Pos 3, calc. or intermediate
G = [WS] Wind Speed, in knots (kt)
H = [WD] Wind Direction, in dec. deg. (T)
I = [TAS] True Air Speed, in knots (kt)
J = [TH] True Heading, in dec. deg. (T)
K = [GS] Ground Speed, in knots (kt)
L = [RLC] Rhumbline Course, 1 to 2 (T)
M = [RLD] Rhumbline Distance, 1 to 2 (nm)
N = [GCD] Great Circle Distance, 1 to 2 (nm)
O = [GCC1] Great Circle Course, at 1 (T)
P = [GCC2] Great Circle Course, at 2 (T)
Q = Mercator Lat. conversion factor, Pos 1
R = Mercator Lat. conversion factor, Pos 2
S = [LOD] Lon. Displ., 1 to 2, in dec. deg.
T = [ET] Elapsed Time, in dec. hours (hr)
U = Utility function
V = Utility function
W = Utility function
X = Utility function
Y = Utility function
Z = Utility function
```

EQUATIONS

Note: Although the equations are presented in a format recognizable to BASIC users, to use them in BASIC, decimal degrees [DEC] must first be converted into radians [RAD], as noted below, and trigonometric functions must be indirectly approximated.

CONSTANT:

01. [PI] Ratio of a circle's Circumference to its Diameter:

$$PI = 3.14159265$$

TEMPERATURE CONVERSION:

02. [X] Fahrenheit (#F); from [Y] Celsius/Centigrade (#C):

$$X = (Y*9/5)+32$$

03. [Y] Celsius/Centigrade (#C); from [X] Fahrenheit (#F):

$$Y = (X-32)*5/9$$

04. [Z] Kelvin (#K); from [Y] Celsius, or Centigrade (#C):

$$Z = Y+273$$

05. [Y] Celsius, or Centigrade (#C); from [Z] Kelvin (#K):

$$Y = Z-273$$

DISTANCE CONVERSION:

06. [Y] Statute Miles (sm); from [X] Nautical Miles (nm):

$$Y = X*1.151594$$

07. [X] Nautical Miles (nm); from [Y] Statute Miles (sm):

$$X = Y/1.151594$$

LATITUDE/LONGITUDE CONVERSION:

08. [V] Latitude or Longitude in decimal degrees (DEC); from [W, X, Y, Z] Latitude or Longitude (DMS), where W = DEG; X = MIN; Y = SEC; and Z = +/-1, (North, East = +1):

$$V = (W+(X/60)+(Y/3600))*Z$$

(Then let A, B, C, D, E, or F = V)

09. [U] Latitude or Longitude in radians (RAD); from [V], Latitude or Longitude (DEC):

$$U = V*PI/180$$

(Then let A, B, C, D, E or F = U)

10. [V] Latitude or Longitude in degrees (DEC); from [U], Latitude or Longitude (RAD):

$$V = U*180/PI$$

11. [W,X,Y,Z] Latitude or Longitude in degrees (DMS); from [V], Latitude or Longitude in degrees (DEC):

V = Latitude or Longitude in degrees (DEC)
Z = ABS(V)*3600 (Total seconds of arc)
Y = INT(Z) (whole or integer seconds of arc)
U = Z-Y (Decimal fraction remainder of seconds)

$$W = INT(ABS(V)) \text{ (DEG)}$$

Z = Y- (W*3600)
(Total min. and sec., in seconds)

$$X = INT(Z/60) \text{ (MIN)}$$

(1) Y = Z-(X*60) (SEC)
(2) Y = Z-(X*60)+U (SEC)

$$Z = SGN(V) \text{ (SIGN, + or -)}$$
("+1" = North/East; "-1" = South or West)

12. [Q] Mercator Conversion Factor (LAC1); for [A] Position 1, Latitude in degrees [DEC]:

$$Q = LNTAN(45+A/2)$$

13. [R] Mercator Conversion Factor (LAC2); for [C] Position 2, Latitude in degrees [DEC]:

$$R = LNTAN(45+C/2)$$

14. [W] Mercator Conversion Factor (LAC3); for [E] Intermediate Position 3, Latitude in degrees [DEC]:

$$W = LNTAN(45+E/2)$$

15. [S] Longitudinal Displacement (LOD1); from Position 1 (B) to Position 2 (D), Longitude in degrees [DEC]:

$$S = ACS(COS(D-B))*SGN(SIN(D-B))$$

16. [S] Longitudinal Displacement (LOD2); from Position 1 (A) to Position 2 (C), *Latitudes* (A & C) in degrees [DEC], and Course (L) from Position 1 to 2, in degrees [DEC]:

$$S = TANL*(R-Q))*180/PI$$

17. [S] Longitudinal Displacement (LOD3); for Position 1 (A) to Intermediate Position 3 (E); *Latitudes* (A & E) in degrees [DEC]; Course (L), in degrees [DEC]:

$$S = TANL*(LNTAN(45+E/2)-Q)*180/PI$$

18. [Z] Corrective Factor (COR1); for resolving calculated bearing lines, from S (from Equation 15, 16 or 17):

$$Z = 90*(1-SGN(S))$$

19. [Y] Latitude Plot (LAP); for plotting position Latitude, where C = Position Latitude; A = Map Origin Latitude; W = Map Scale; Latitude in degrees [DEC], Scale in in/deg:

$$Y = (LNTAN(45+C/2)-LNTAN(45+A/2))*W *180/PI$$

20. [X] Longitude Plot (LOP); to plot position Longitude, where D = Position Longitude; B = Map Origin Longitude; W = Map Scale; Latitude in degrees [DEC}, Scale in in/deg:

$$S = ACS (COS(D-B))*SGN(SIN(D-B)) \text{ (Equation 15)}$$

$$X = ABS(S)*W$$

TIME CONVERSION:

21. [T] Elapsed Time in decimal hours; from [U, V, W, X, Y, Z], variable function GMT Start and End data, where:

U = Start Date in month: V = Start Hour: W = Start Minute
X = End Date in month: Y = End Hour: Z = End Minute

$$T = ((X*24)+Y+(Z/60))-((U*24)+V+ (W/60)) \text{ (same month)}$$
$$T = (Y+(Z/60))-(V+(W/60)) \text{ (same day)}$$

WIND TRIANGLE EQUATIONS:

22. [G] Wind Speed (WS); in knots (kt):

$$G = SQR(K^2+I^2-2*K*I*COS(J-L))$$

23. [H] Wind Direction (WD); in degrees True (#T):

$$H = J-ACS((I-(K*COS(L-J)))/G)$$

24. [I] True Air Speed (TAS, TS); in knots (kt):

$$I = SQR(G^2+K^2-2*G*K*COS(L-(H-180)))$$

25. [J] True Heading (TH); in degrees True (#T):

$$J = L+ASN(G*SIN(H-L)/I$$

26. [K] Ground Speed (GS); in knots (kt):

$$K = I*COS(J-L)-G*COS(H-L)$$

27. [L] True Course (TC); in degrees True (#T):

$$L = J + ASN(G*SIN(J-H)/K)$$

RHUMBLINE EQUATIONS:

28. [L] Rhumbline Course (RLC) in degrees (DEC); from Position 1 (A) to Position 2 (C), Latitude and Longitude in degrees [DEC]:

$$S = ACS(COS(D-B))*SGN(SIN(D-B)) \text{ [Equation 15]}$$
$$Z = 90*(1-SGN(S)) \text{ [Equation 18]}$$

$$L = -ATN(180*(R-Q)/PI/S) + 90 + Z$$

29. [M] Rhumbline Distance (RLD1) in Nautical Miles (nm); from Position 1 to Position 2, data in degrees [DEC], for all course values *except* due-North, -East, -South and -West:

$$M = (C-A)*60/COSL$$

30. [M] Rhumbline Distance (RLD2) in Nautical Miles (nm); only for courses due-East or due-West, from Position 1 (A,B) to Position 2 (D), position data in degrees [DEC]:

$$S = ACS(COS(D-B))*SGN(SIN(D-B)) \text{ (Equation 15)}$$

$$M = S*COS(A)*60$$

31. [M] Rhumbline Distance (RLD3) in Nautical Miles (nm); only for courses due-North or due-South, from Position 1 (A) to Position 2 (C), Latitudes in degrees [DEC]:

$$M = (C-A)*60$$

32. [E] Rhumbline Intermediate Latitude (RLE1), in degrees [DEC]; from Position 1 Latitude (A), Course (L), Distance (M) in Nautical Miles (nm); other data in degrees [DEC]:

$$E = A + M*COSL/60$$

33. [E] Rhumbline Intermediate Latitude (RLE2), in degrees [DEC]; from Position 1 Longitude (B), Intermediate Position Longitude (F), and Course (L), all data in degrees [DEC]:

$$X = ACS(COS(F-B))*SGN(SIN(F-B))$$

$$E = 2(ATN(EXP(Q-(PI*X/180/TANL)))-45)$$

34. [F] Rhumbline Intermediate Longitude (RLF1), in degrees [DEC]; from Position 1 Longitude (B), Intermediate Latitude (E) and Course (L), all data in degree [DEC]

$$Y = TANL*(LNTAN(45+E/2)-Q)*180/PI$$

$$F = ACS(COS(Y+B))*SGN(SIN(Y+B))$$

35. [F, E] RhumbLine Intercept Position (RINT) of two RL Courses; in degrees [DEC]; from Position 1 (A,B) and Position 2 (C), with all Position data in degrees [DEC]:

U = RhumbLine Course (Interceptee); from Position 1.
V = RhumbLine Course (Interceptor); from Position 2.

$$X = SIN(V-L)*SQR(((Q-R)*180/PI)^2+S^2)$$
$$*COS(U+90)/SIN(V-(U-180))$$
$$Y = ACS(COSX)*SGN(SINX)$$

$$F = ACS(COS(B+Y))*SGN(SIN(B+Y))$$
(Intercept Longitude)

$$E = 2*(ATN(EXP(Q-(PI*(-F)/180/TANQ)))-45)$$
(Intercept Latitude)

Note: Although the terms "Interceptee" and "Interceptor" have been used, the applications are interchangeable.

GREAT CIRCLE EQUATIONS:

36. [N] Great Circle Distance (GCD) in Nautical Miles (nm); from Position 1 (A,B) to Position 2 (C,D):

$$N = ACS(SINC*SINA + COSC*COSA*COSS)*60$$

37. [O] Great Circle Initial Course (GCCI) in degrees (DEC); from Position 1 (A,B) to Position 2 (C,D), with all data in degrees [DEC]:

$$O = ACS((SINC-SINA*COS(N/60)) /COSA/SIN(N/60))*SGN(S)+2*Z$$

38. [P] Great Circle Final Course (GCCF) in degrees (DEC): from Position 1 (A,B) to Position 2 (C,D), with all data in degrees [DEC]:

$$P = ACS((SINA - SINC*COS(N/60))/COSC/SIN(N/60))*SGN(-S)+180$$

39. [E] Great Circle Intermediate Position Latitude (GCI1); in degrees [DEC], from Position 1 (A,B) to Position 2 (C,D) and Position 3 Longitude (F), all data in degrees [DEC]:

$$S = ACS(COS(D-B))*SGN(SIN(D-B)) \text{ (Equation 15)}$$
$$X = ACS(COS(F-B))*SGN(SIN(F-B))$$
$$Y = ACS(COS(F-D))*SGN(SIN(F-D))$$

$$E = ATN((TANC*SIN(X)-TANA*SIN(Y)))/SINS$$

40. [E] Great Circle Intermediate Position Longitude (GCI2) in degrees [DEC]; from Position 1 (A) and Bearing 1 [O], in degrees [DEC] and Distance [N] in Nautical Miles (nm):

$$E = ASN(SINA*COS(N/60) + COSA*SIN(N/60)*COSO)$$

41. [E] Great Circle Intermediate Position Latitude (GCI3) in degrees [DEC]; from Position 1 (A), Bearing 1 (O) and Bearing 2 [P], all data in degrees [DEC]:

$$E = ACS(SINO*COSA/SINP)$$

42. [F] Great Circle Intermediate Position Longitude (GCI4) in degrees [DEC]; from Position 1 (A,B) and Intermediate Position 3 (E), in degrees [DEC] and Distance (N), in Nautical Miles (nm):

$$Y = ACS((COS(N/60)-SINE*SINA)/(COSE*COSA))$$

$$F = ACS(COS(B+Y))*SGN(SIN(B+Y))$$

43. [E, F] Great Circle Intercept Position (GINT) of Great Circle Bearings; from Pos 1 and Pos 2, in degrees [DEC]:

U = Position 1 Bearing, in decimal degrees (DEC).
V = Position 2 Bearing, in decimal degrees (DEC).

$$W = O-U$$
$$X = V-(P-180)$$
$$Y = TAN((N/60)/2)$$
$$Z = ATN(COS((X-W)/2)*Y/COS((X+W)/2)+ATN(SIN((X-W)/2)*Y/SIN((X+W)/2)$$

$$E = ASNA*COSZ+COSA*SINZ*COSU$$
(Intercept Latitude)

$$X = ASN(SINQ*SINZ/COSE)$$
$$Y = ACS(COSX)*SGN(SINX)$$

$$F = ACS(COS(B+Y))*SGN(SIN(B+Y))$$
(Intercept Longitude)

Note: GC Intercept equations are used for RDF Bearings.

DYNAMIC FLIGHT EQUATIONS:

44. [M] Distance Made Good (DMG) in Nautical Miles (nm); from Speed Made Good (SMG) [K] in knots (kt) and Time Made Good (TMG) [T] in Decimal Hours (hr):

$$M = K*T$$

45. [K] Speed Made Good (SMG) in knots (kt); from Distance Made Good (DMG) [M] in Nautical Miles (nm) and Time Made Good (TMG) [T] in Decimal Hours (hr):

$$K = M/T$$

46. [T] Time Made Good (TMG) in Decimal Hours (hr); from Distance Made Good (DMG) [M] in Nautical Miles (nm) and Speed Made Good (SMG) [K] in knots (kt):

$$T = M/K$$

INSTRUMENT EQUATIONS:

47. [X] Approximate Standard Mean Temperature at given True Altitude (SMTA); in Degrees Centigrade (#C):

V = True Altitude (TA): in feet above Sea Level

$$X = 15-V/1000-(((V/10000)^2)/10)$$

48. [U] Indicated Altitude (IA), in feet (ft); from the desired, or given, True Altitude

V = True Altitude (TA); in feet (ft) above Sea Level.
Z = True Temperature at Sea Level (TTSL); degrees C.
W = True Temperature at True Altitude (TTTA); degrees C.
X = Standard Mean Temperature specified at True Altitude.
(approximately 5.9C @ 9100ft)

$$U = V*(X+273)/(((Z+W)/2)+273)$$

49. [Y] True Pressure (TPTA) in inches of Mercury (in Hg); at True Altitude.

Z = True Temperature at Sea Level (TTSL); degrees C.
W = True Temperature at True Altitude (TTTA); degrees C.
X = True Pressure at Sea Level (TPSL); (in Hg).
V = True Altitude (TA): in feet (ft) above Sea Level

$$Y = X/10^{\wedge}(V/(221.152*((Z+W+(273*2))/2)))$$

50. [Z] Indicated Air Speed (IAS1) in knots (kt); roughly corrected for True Altitude in feet (ft):

U = Indicated Altitude in feet (ft)
I = True Air Speed (TAS) in knots (kt)

$$Z = I/(1+U*.02/1000)$$

51. [Z] Indicated Air Speed (IAS2) in knots (kt): accurately corrected for True Altitude

I = True Air Speed (TAS) in knots (kt)
W = True Temperature at True Altitude (TTTA); degrees C.
Y = True Pressure at True Altitude (TPTA); (in Hg).

$$Z = I/SQR((W+273)*29.921/288/Y)$$

DERIVED FUNCTIONS FOR BASIC:
(for Inverse Functions which are not directly supported)

52. Inverse Sine (ASN) of (X), or ARC SINE(X):

$$ASN(X) = ATN(X/SQR(-X*X+1))$$

53. Inverse Cosine (ACS) of (X), or ARC COSINE(X):

$$ACS(X) = -ATN(X/SQR(-X*X+1))+1.5708$$

MATHEMATICAL OPERATORS

1. X+Y, X-Y, X*Y, X/Y: X plus Y, X minus Y, X times Y, X divided by Y

2. X=Y, X>Y, X<Y: X is equal to, greater than, less than, the value of Y

3. SIN(X), COSX, TANX: the Sine, Cosine, Tangent functions: of X or (X)

4. ASNX, ACS(X), ATNX: the Arcsine, Arccosine, Arctangent: of X or (X)

5. SQR(X), X^2: Square Root of (X), X to the second power (squared)

6. LN(X), EXP(X): Natural Logarithm of (X), Inverse Logarithm of (X)

7. ABS(X): The absolute (positive) value of the numerical expression (X)

8. SGN(X): The sign, + or -, of (X), expressed, respectively, as +1 or −1

9. INT(X): Defines (X) as the largest integer or whole number equal to or less than (X); as in, INT(5.6) = 5, or, INT(-5.6) = -6.

INTERDEPENDENT EQUATIONS: The list below refers to the general equations listed above. The required executions must be performed in the order noted. In some cases, other designators may be used. Every effort has been made to assure universality and accuracy of application.

01. 02 through 11, and 21 may be executed independently.

02. 12 through 15 require execution of 08; twice for each position.

03. 16 and 17 require execution of 08, 12, 13 and 28.

04. 18 requires execution of 15, 16, or 17.

05. 19, 20 require execution of 08 and 15.

06. 22 through 27 are interdependent or may require data entry.

07. 28 requires execution of 08,12,13,15,18.

08. 29 requires execution of 08, 28.

09. 30 requires execution of 08, 15.

10. 31 requires execution of 08.

11. 32 requires execution of 08, 28, 29 or a specific data entry.

12. 33 requires execution of 08, 12, 28, and data entry (F).

13. 34 requires execution of 08, 12, 28, and data entry (E).

14. 35 requires execution of 08, 35, and data entry (E).

15. 36 requires execution of 08, 15.

16. 37 requires execution of 08, 15, 18, 35.

17. 38 requires execution of 08, 15, 35.

18. 39 requires execution of 08, 15.

19. 40 requires execution of 08, 35, 36.

20. 41 requires execution of 08, 36, 37.

21. 42 requires execution of 08, 12, 13, 15, 28, and data entry.

22. 43 requires execution of 08, 36, 37, 35, and data entry.

23. 44 through 46 are interdependent.

24. 47 through 51 are interdependent.

25. 52 and 53 are independent.

5. AFTERWORD: Opposing Forces—Reflections of the Author

Section 2.2 presents the Philippine Sea area, scene of the *Hawaii Clipper* hi-jacking, as a "Western Pacific Game-board," the players being, of course, the respective navies of the Empire of Japan and of the United States. The Clipper, a pawn in that game, continues to be a game-piece, but in a different game, today, one where the governments of Japan and the United States play as partners on one side of the board, and take on challengers, such as Jackson, Gervais, and myself, on the other. Until recent years, I had wondered, as Jackson and Gervais must have wondered, why the search for the truth about *Hawaii Clipper* should even be a game at all, and why Washington has joined with Tokyo to play such a serious game of "hide-and-seek" over this incident: a true story of historical significance—and a yarn worth spinning.

Although it had been evident, for years, that a post-war *black accord* existed between the United States and Japan, it was not until the morning of June 14, 1994, that the virulence of this accord was made abundantly clear. It was during that morning's news broadcasts that President Clinton's unpublicized state dinner of the previous evening, honoring the Emperor of Japan, was first made public. (See page ix of this work. For those readers who require more detailed information on this state dinner, see *The New York Times* of June 14, pages A, 1:3 and B, 8:1 and 8:4, as well as June 19, page IV, 4:3, for a story on the President's *bow* to the Emperor. There was *no* prior announcement of this affair.)

National memorials include *dates*, as well as statues and monuments. This was certainly made clear in the official U.S. recognition of the fiftieth anniversary of the June 6, 1944, invasion of Normandy, which heralded the beginning of the end for Nazi Germany. The German vets, incidentally, were specifically excluded from the 1994 memorial, as they had *not* been, in 1984. It will be remembered that the President was a highly visible figure at Normandy, on June 6, 1994, and that he had attended a dinner the previous evening, June 5, across the channel in England, where he dined with U.S. veterans of the invasion.

But the invasion of Saipan, on June 15, 1944 (June 14, 1944, in the U.S.), was far more significant for Americans: in 1944, Germany had no immediate means by which to attack the United States, but Japan was even then prepared and planning to initiate a first wave of direct attacks on America. The first wave of balloon attacks began in November, 1944, and continued through May, 1945. This first wave was unsuccessful, but it paved the way for the second wave, which was to begin in the fall of 1945, making use of the 200mph Jet Stream to carry plague and anthrax to the American heartland. *Unlike Normandy, the U.S. victory at Saipan had a direct effect upon the <u>survival</u> of America and its People.*

Yet, Saipan was all but forgotten in June of 1994: There was a blurb in *The Cincinnati Enquirer*, a column inch in *The Washington Post*, nothing in *The New York Times* and *nothing* in the national broadcast media: the President's silence on Saipan had involved a *media blackout*, as well as a deliberate snub of the Saipan invasion. But, in honoring the Emperor of Japan at the White House *on that date*, with no American veterans of the invasion present, the President had provided *aid and comfort*, if not to a present enemy, then to the enduring *spirit* of past enmity.

The President committed an Act of Treason, and desecrated a national memorial. The American People and their Congress should have been outraged, but fear rules America, and the "partners" suppress its history.

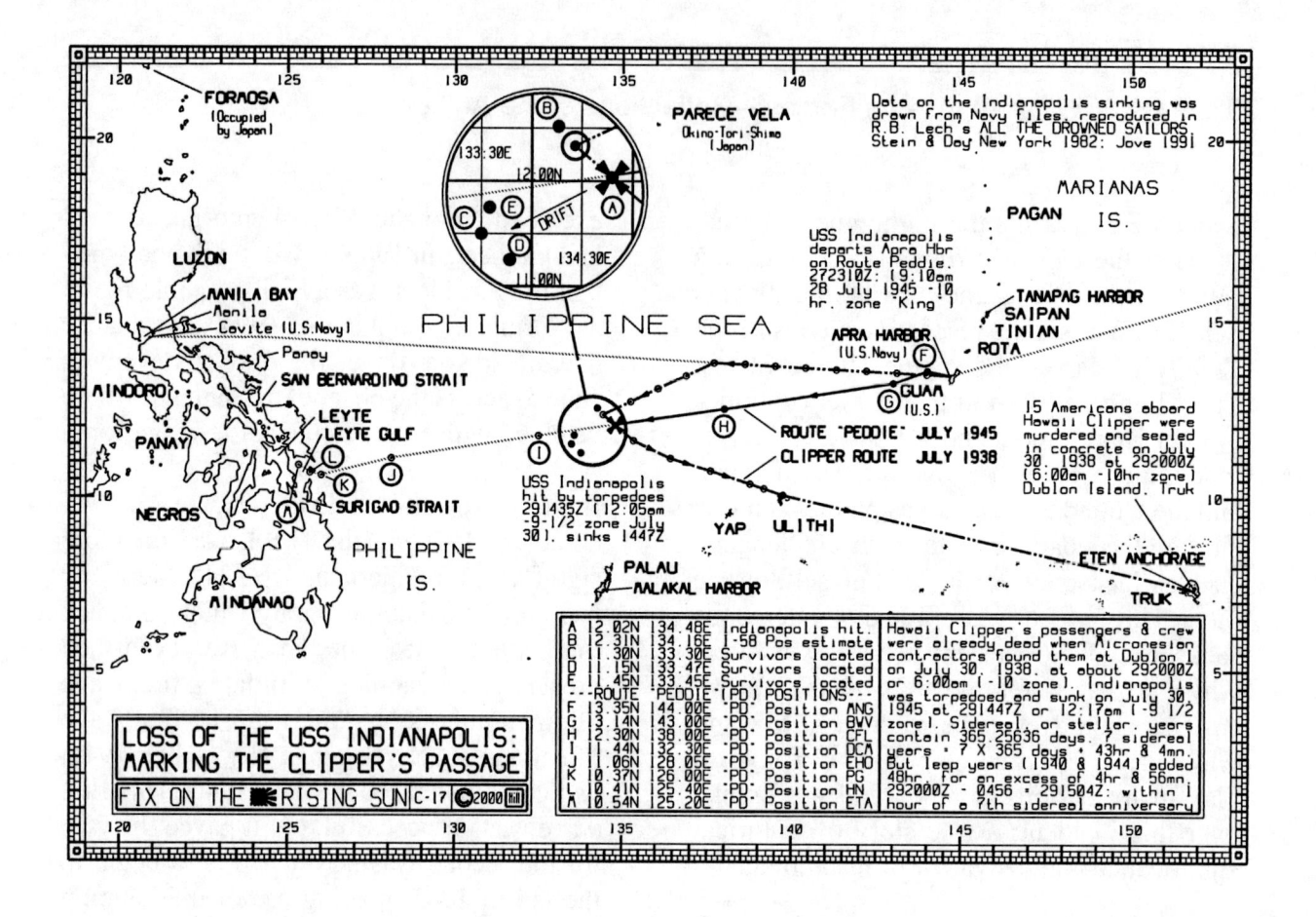

It may appear that a challenger would stand no chance against such adversaries, but it is my belief that we don't stand alone. Jackson was driven by a force that he could not explain, and I, too, have pressed this issue beyond reasonable limits. However, while attempting to identify this "driving force," I came across a singular coincidence:

When the cruiser *Indianapolis* was sunk by torpedoes in 1945, it was the worst "at-sea" disaster in the history of the U.S. Navy and, also, the first instance of an enemy officer being called to testify against the captain of an American warship. At his court martial, Captain Charles McVay testified that, on the night of 29/30 July, 1945, *Indianapolis* had been proceeding, *at 17kt*, along Route PD, from Guam to Leyte Gulf, when she endured two explosions and sank. Cdr. Mochitsura Hashimoto, captain of the Imperial Japanese submarine, *I-58*, testified that he had fired a six-torpedo spread—*using a target speed of 12kt!* Believing that *four* torpedoes had hit *Indianapolis*, he revised this target speed to *11kt*, for his report. However, as only *two* had actually hit, *Indianapolis'* speed must be further revised—*down to 10kt!* McVay had, for some reason, reduced speed to *only 10kt*, at about 291315Z, and later—*lied* about it. That's why Hashimoto had been brought in.

What "driving force" could have put such a thought into McVay's mind, that he would be secure at only 10kt? Perhaps he could not sleep, tossing on "the confused sea," through which *Indianapolis* was plowing, yet, had he maintained 17kt, he would have passed *I-58* before Hashimoto awoke and surfaced. But he gave in to—*something*—and *Indianapolis* went down just as she was about to cross the Clipper's route to Ulithi, at 291447Z, on 30 July, 1945—seven *sidereal* (or navigator's) years, *almost to the hour*, after the Clipper's passengers and crew were murdered. And what cause have *they* had, to *rest in peace*?

6. INDEX

A

Aero Club of Portland....................................4
Aero-Biology
    Air-Hook..6, 7
    Committee on Aero-biology 6
    Japanese Balloon Assault....................7, 177
    Japanese Plague Development..............7, 177
    Jet Stream, Japan to America...............7, 177
    Leprosy 5, 6
    Micro-organisms..............................xxv, 6
Agana, Guam.................................41, 45, 170
Air Link (Japan/PAA)................60, 81, 92, 94, 101
Air Safety Board (CAA)............. v, xiv, xvii, xviii, xx, 12, 47, 49, 53, 55, 57, 77, 78, 86, 89, 90, 91, 110, 117, 118, 120, 121, 122, 123, 124, 125, 126, 127, 128, 129, 130, 131, 132, 133, 134, 135, 136
    Preliminary Report .... xiv, xvii, xviii, xx, 47, 49, 53, 55, 77, 78, 86, 89, 90, 91, 110, 117, 118, 120, 121, 122, 123, 124, 125, 126, 127, 128, 129, 130, 131, 132, 133, 134, 135, 136
Amelia Earhart Research Consortium .... xi, xii
American Brass Company 10
American Express.....................................4
American Samoa.............................xxvi, 18, 46
American Volunteer Group xxii
Annapolis, Maryland...............................xi
Anthony, David xii
Apra Harbor, Guam.................35, 41, 47, 48, 49, 53, 54, 58, 59, 73, 79, 95, 96, 170
Arey, James A.................................xiv, xx, 70
Arnold, Patrick ....xx, 110
Asia.........................3, 4, 6, 27, 37, 39, 41, 46
Aslito Airfield, Saipan..............................65
Associated Press ....xx, 60, 70, 75, 110, 142
Atlanta, Georgia.......................xx, 86, 110, 113
Atlantic ....xxiv, 6, 17, 22, 24
Australia ....xxvii, 14
Ayres, Frank, Jr. xiii

B

Backus, Jean L....................................... xiii
Baker Island...165
Balfour, Harry J. ....xiv, xxiv, xxvi, 156, 157
Baltimore, Maryland................................39
Barrett Airways..18
Bellarts, Leo G.................xiv, xxv, 156, 158, 161
Belotti, Ellen..................................xxv, 31
Bendix Corporation
    Blind Landing System..............................56
Benzon, Robert xii
Berlin-Tokyo (-Rome) Axis .... xxiii, xxiv, xxvi, 27

Bertrandias, Neil..4
Bismark Archipelago................................37
Black Sea..71
Black, Richard Blackburn..................... xii
Boeing
    314 (Flying Boat Airliner) .... xxviii, 17, 22, 24
Bolsheviks 18
Bongard, David L. xiii
Brazil..18
Bridgeport, Connecticut..........................106
Bronxville, New York 9
Buffalo, New York 9, 10
Bulolo, New Guinea xxvi
Burbank, California 56

C

Canacao Bay, P.I. 45, 51, 170
Canada...7, 10
Canaday, Harry......................xii, 45, 46, 51
Canton, China 3, 4, 5, 29, 84, 91
Caribbean.............................12, 17, 18, 39
Caroline Islands.............................36, 37, 86
Carondelet reef 165
Carripito, Venezuela..............................23
Catanduanes Island, P.I. 55
Cavite, P.I. .... xviii, 25, 35, 36, 39, 40, 43, 44, 45, 46, 47, 48, 50, 51, 53, 55, 57, 59, 63, 68, 71, 73, 75, 77, 81, 91, 96, 97, 98, 99, 100, 101, 170, 171
    Landing Area 45, 59
Central America.....................................17
Charan Kanoa, Saipan 65
Chevy Chase, Maryland............................6
China Incident.......................................28
China Lake, California............................56
China National Aviation Corporation (CNAC).........5, 10, 27, 46, 91, 94, 106
    Kweilin.................5, 27, 29, 91, 94, 106
Chinese Air Force...............................3, 92
Choy, Wah Sun..................................... xxii, 3, 5, 10, 66, 83, 84, 92, 93, 138
Chungking, China..................................3, 5
Church, David....................................... xii
Cipriani, Frank.......................................158
Civil Aeronautics Authority (CAA) .... v, xiv, xvii, xviii, xxiv, 49, 57, 77, 90, 110, 117, 118, 119, 120
Clinton, British Columbia.......................10
Clinton, William Jefferson.................ix, 66, 177
Cohen, Stan... xiii
Cold War xxvi
Collopy, James A...................................xiv

Colonialism..37
Columbia University....................................18
Concrete Entombment Slab ....v, xx, xxi, xxii, 36, 65, 84, 85, 86, 89, 93, 102, 106, 109, 149
Coop, Elizabeth McCarty 18
Cooper, Dan..xxvi
Corregidor, P.I...14
Corrigan, Douglas (*Wrong Way*) xxiv
Cox, Howard L. 17, 18, 100
Cuba..19, 107
Curtiss-Wright Corporation 9, 10, 18, 35, 92
   Hawk 75 (Pursuit) 9, 10, 83, 84, 92
   Hawk 75-A (Pursuit) 10, 83, 84, 92
   P-36 (Pursuit).................................10, 83

D

Daley, Richard...xiii
Daley, Robert... xiii, xix, xxi, 12, 19, 22, 83, 105, 107
Davis, George M.............................17, 19, 21, 24, 36, 63, 64, 67, 68, 75, 76, 79, 80, 96, 97, 98, 99, 100
Devine, Thomas E. ....xiii, 65
Direct Route...................39, 40, 41, 42, 43, 44, 51, 55, 58, 59, 73, 75, 97, 167
Dole Air Race, 1927.....................................xxiv
Douglas Aircraft Co......................................17
   DC-2 (Airliner) 5, 27, 91, 106
Dublon Island, Truk Atoll...................v, xii, xx, 36, 69, 70, 84, 85, 89, 93, 101, 109, 110, 148, 170
Dwiggins, Don...xiii

E

Earhart, Amelia Mary .... xi, xii, xiii, xiv, xx, xxii, xxiv, xxv, xxvi, xxvii, 6, 7, 12, 21, 22, 23, 24, 27, 29, 30, 31, 40, 48, 65, 70, 85, 86, 94, 109, 155, 156, 157, 158, 159, 160, 161, 162, 163, 164, 165, 187
   American flying lady......................63, 65
   Foreign Woman 30, 31
   Last Flight..................................... xiii, 6, 23
   LOP 157-337 ....xxiv, 155, 159, 161, 164, 165
   Running north and south....................xxiv, 159, 162, 163, 164, 165
   Russian Woman................................30, 31
   Soviet Communism..........................xxiv, xxvi, 31
   Twice A Traitor 31
Eisenhower, Dwight David...........................14
Ellis, Earl 61, 62, 85
Emperor of Japan
   Akihito ....ix, 66, 177
   Hirohito ix
Engines, Aircraft........................... xix, xxi, xxii, xxv, 10, 12, 19, 23, 35, 53, 69, 74, 82, 83, 84, 89, 92, 95, 101, 102, 107
Eten Anchorage, Truk........................67, 98, 99, 101
Ethell, Jeffrey L. xiii
Etorofu Island, Kuriles.................................37
Evanston, Illinois...10

Evidence Summary
   01. The Rumors...89
   02. The Clipper 89
   03. The Traces 89
   04. The Navigation Analysis....................90
   05. The Oil Slick 91
   06. The Lahuy Report..............................91
   07. The Flight Reports.............................91
   08. The Passengers..................................91
   09. The Search 92
   10. The Japanese Air Route 92
   11. The Political Games..........................93
   12. The Nauroon/Mori Allegations........93
   13. The Acts of Japanese Aggression 94
   14. The *H6K* Support Flight...................94
Eymann, Eleanor (Jewett).....................xi, 12, 17, 30

F

Federated States of Micronesia........xi, xxi, 36, 69, 86
Fernandez, Edouard.....43, 50, 63, 64, 67, 80, 99, 100
Flying Tigers (AVG) ....xxii, 35, 83, 84, 92
Formosa 60, 73, 89, 97, 101, 170
Forrestal, James 65
Fort McKinley, P.I..45
Fort Ticonderoga, New York.......................14
France ....xxiii, 37
Freeman, Richard..xi
French, Howard C. 3, 4, 10, 84, 92
Fuchida, Mitsuo 27, 28, 29
   Christianity, conversion to 27

G

Galten, William L. ....xiv, 158, 161, 162, 163
Garapan City, Saipan..................................65
Gardner Island (Nikumaroro), Phoenix Is. 165
Gasoline....xxvi, 5, 46, 47, 53, 57, 73, 91, 94, 99, 106
Genda, Minoru............................... vii, xii, 27, 29, 31
   Madman Genda................................27
George Washington University 5
Germany .... xxiii, 37, 71, 177
   Nazis .... xxiii, 7, 27, 177
Gervais, Joseph L. .... v, vii, xi, xii, xiii, xiv, xx, xxi, xxii, 14, 65, 67, 70, 85, 86, 93, 105, 109, 110, 113, 142, 143, 148, 149, 151, 153, 154, 177, 187
Gifu, Japan..69
Goerner, Frederick......................................xiii
Graham, Roberta..xi, 5
Gray, Almon A. ....xii, xiii
Great Britain ....xxiii, 10, 37
Great Depression 17, 22, 24, 83
Greene, Fred xiii
Greenwood, Robert B. .... xi, xii, xiv, xxi, xxviii, 19, 21, 22, 23, 24
Guam ....v, xxii, 9, 18, 23, 35, 37, 39, 41, 42, 43, 45, 47, 49, 50, 53, 54, 59, 60, 63, 64, 67, 71, 74, 75, 77, 81, 90, 91, 95, 96, 100, 101, 102, 107, 170, 171, 178

Gwynne-Jones, Terry...xiii

H

Hainan Island, China 29
Hankow, China 29
Hardin, Thomas O. xiv
Harvard University 6
Harvey, Ralph 21
Hashimoto, Mochitsura.............................178
Hawaii Clipper
  1. Captain........................51, 54, 79, 91, 96
  2. First Officer............. xii, xxi, xxv, xxviii, 17, 19, 21, 22, 24, 35
  3. Second Officer 17, 21, 24, 36, 51, 63, 64, 79, 80, 96, 97, 98
  4. Third Officer ....xxii, 18, 24, 98
  5. Fourth Officer 12, 17, 98
  6. Engineer Officer 17, 18, 51, 100
  7. Assistant Engineer Officer 17, 18, 95
  8. Flight Radio Officer............................18, 35, 36, 51, 52, 63, 64, 89, 98
  9. Flight Steward................................18, 95
  Ground Speed (Final)................50, 57, 171
  Ground Speed (Initial) 54, 58, 75, 79, 90
  Maximum Range.............................47, 91
  True Air Speed (Optimum).....................47, 48, 50, 53, 54, 57, 58, 67, 71, 73, 75, 90
Hawaiian Islands....................................5, 65
Hezel, Francis xi
Hi-jacking .... xi, xii, xvii, xix, xx, xxi, xxv, xxvi, 21, 24, 25, 27, 29, 31, 33, 59, 60, 61, 62, 64, 66, 67, 68, 69, 71, 72, 73, 74, 75, 79, 82, 83, 84, 89, 91, 92, 94, 97, 99, 100, 105, 106, 107, 110, 167, 177, 187
  Course D-282T 57
  Course D-Cavite Landing 57
  Course D-E EXT.......................50, 51, 52, 57
  Course D-E1 EXT...............................57, 73
  Course E-282T 50, 51, 52
  Lahuy Overflight..................42, 55, 78, 102
  Leg A-B 50, 54, 58, 67, 75, 91
  Leg A-D Direct 58, 76, 90
  Leg B-C1 54
  Leg C1-D 54
  Leg D-E 50, 51, 52, 54, 57, 59, 63, 68, 90, 91, 100, 170
  Leg D-E1 57
  LOP A-180T (Palau pointer) 63, 64, 67, 170
  LOP A-B1 (Truk pointer) 67, 170
  LOP C-A (Saipan pointer) 63, 64, 67, 68, 170, 173
  LOP D-E (Cavite pointer)...........57, 59, 63, 68, 90
  LOP Pointers...............59, 63, 67, 68, 75, 79, 165
  Magneto Serial Number........ xix, xxii, 83, 89, 107
  Math Match..58
  Minor Hitch..79
  Oil Slick.. xviii, xxvii, 53, 57, 68, 73, 76, 77, 78, 81, 91, 92, 94, 99

Original Hi-jack Plan.......59, 67, 74, 75, 76, 97, 99
Philippine Telephone Co...........53, 55, 76, 78, 101
Position A 49, 50, 53, 58, 59, 62, 63, 67, 71, 73, 75, 76, 79, 80, 81, 90, 91, 170
Position A1 ....58, 59, 62, 63, 67, 71, 73, 75, 76, 79, 170
Position B.......................49, 54, 57, 63, 64, 67, 68, 75, 76, 79, 80, 90, 91, 170
Position B0 ....75, 76, 79, 80, 170
Position B1......................67, 68, 75, 79, 80, 170
Position C................49, 54, 57, 63, 64, 68, 75, 76, 79, 80, 90, 91, 170
Position C0 ....75, 76, 79, 80, 170
Position C1............54, 63, 68, 75, 76, 79, 80, 170
Position D 49, 50, 57, 58, 59, 63, 68, 71, 72, 73, 75, 79, 80, 91, 170
Position D0 ....75, 80, 170
Position D1 ....71, 72, 75, 170
Position E 49, 50, 51, 53, 54, 57, 68, 71, 72, 75, 80, 92, 99, 170
Position E0 ....75, 80, 170
Position E1 ....57, 68, 71, 72, 80, 170
Position E2 ....71, 72, 75, 170
Position F 50, 53, 68, 76, 79, 170
Position F0 170
Position F1 170
Position F2 170
Position Oil Slick (*Meigs*)............. xviii, xxvii, 53, 57, 68, 73, 76, 77, 78, 81, 91, 92, 94, 99, 170
Position Oil Slick Origin................53, 57, 78
Position P8 ....75, 170
Position Ulithi 170
Positions (False)..67
Positions (Reconfigured False) 67
Positions (True).................................67, 90
Rising Sun-Lines .60, 63, 64, 67, 75, 79, 83, 90, 99
Route A-D (Math Match).........................75
Route Apra to A (Direct) 58
Route Apra to Cavite (Standard).................48
Route Apra to D (Direct) 58
Route Apra-A/A1 (Planned) 58
Route Apra-P8-A (Actual)...............53, 75, 170
Hiroshima, Japan 27
Hittokapu Bay, Etorofu Island.....................37
Hong Kong .... xxiv, xxvii, 3, 5, 9, 10, 35, 40, 46, 84, 91, 106, 170
Horne, W. T. 65, 83
Hoshina, Zenshiro.......................................28
Howland Island.....................................xxiv, xxvi, 23, 155, 156, 157, 158, 159, 163, 164, 165
Hoyt, Robert D. ....xiv, 136
Hughes, Howard 29, 30
H-1 (American Racer).........................29, 30

I

Imperial Airways 10, 46
Imperial General Staff 15

Imperial Japanese Navy .... v, xviii, xx, xxi,
   5, 24, 27, 30, 31, 35, 36, 37, 38, 62, 63, 65, 67, 81,
   82, 92, 93, 95, 96, 99, 101, 102, 109
   Fleet Faction 27, 35, 37, 69, 102
   Fourth Fleet .... v, 36, 67, 93, 102, 109
   Hospital (Dublon) .... v,
      85, 86, 93, 102, 109, 110, 149
   Staff College 27, 28, 29
   Thirteenth Naval Air Corps 28
   Treaty Faction 38
Infanta, P.I. 40, 44, 55, 170
International News Service 7
Investigations
   1. The First xviii
   2. The Second xviii
   3. The Third xix
   4. The Fourth xix
   5. The Fifth xx
   6. The Sixth xx
   7. The Seventh xxi
Iran/Contra 31
Ise Bay, Japan 82, 102, 170

J

Jackson, Ronald W. .... xiii, xix, xx, xxi, xxvii,
   18, 21, 22, 59, 60, 61, 86, 89, 105, 106, 107, 110,
   177, 178
Jewett, John W. .... xi, 12, 17, 19, 96, 98
Johnson, Curt xiii
Johnson, J. Monroe 21, 106
Johnson, John L., Jr. xii
Johnston Island 7, 156
Jones, John Paul 37
Jordanoff, Assen .... xiii, 56

K

Katz, Harvey .... xix, 107
Kaucher, Dorothy xiii
Kawanishi .... xxvii, 38, 73, 82, 94, 95, 106
   H6K Mavis (Flying Boat) .... xxvii,
      38, 73, 74, 75, 76, 79, 92, 94, 96, 97, 98, 99,
      100, 101, 106, 170
Kennedy, Kenneth A. .... xi, xiv, 3, 5, 6, 137
Kennedy, Marjory .... xiv, 5
Kenner, Frank 155
King, Ernest J. 105
Kingsford-Smith, Charles xxiv
Kito Butai (attack force) 37
Klass, Joseph .... xiii, 109, 187
Koror, Palau 38, 60, 61, 62, 91, 92, 94, 97, 98, 170
Kuomintang Government, China 3, 92
Kurile Islands 37

L

Lae, New Guinea .... xxiv, xxvi,
   23, 29, 31, 156, 157, 165
Lake Champlain, New York 14

Laoang, P.I. 43, 49, 170
Las Vegas, Nevada .... xii, 70
League of Nations 37, 60, 61, 69
   Japanese Mandates 22, 23, 35, 36, 37, 38,
      59, 60, 61, 62, 63, 69, 73, 101, 165
Lee, Mary Ann (Walker) .... xi, xxi,
   18, 19, 21, 24, 25, 26
Lee, Travis xi
Leuder, Karl 106
Leuteritz, Hugo xii
Lindbergh, Charles 6
Lockerbie, Scotland .... v, 107
Lockheed .... xx, xxv, 65, 161
   10E Electra (Airliner) .... xx, xxv, xxvii,
      23, 65, 85, 155
London, England 9
Long Beach, California 24
Long Island, New York 18
Loomis, Vincent V. xiii
Loree, Andy xi
Lueke, Lois 21
Luttrell, John F. .... xi, xx, xxi, 86, 110, 113
Luzon, P.I. 43, 49, 51, 53, 55, 56, 73, 91, 100, 101

M

Macao, China 46
MacArthur, Douglas .... xi, xii, xix, 14, 69, 83
Mack Truck Company 12
Magley, Dean xii
Malakal Harbor, Palau 59, 63, 73, 98, 170
Manila Hotel 45
Manila/Manila Bay .... v, xxiv, xxvii,
   5, 23, 25, 35, 37, 41, 43, 44, 45, 46, 49, 51, 53, 55,
   76, 77, 78, 101, 170
Mantz, Paul 30
Marehalau, Jesse xi
Marianas Islands 37, 102, 146, 170
Marpi Point, Saipan 65
Marshall Islands .... xiv, xxv, xxvi,
   30, 31, 37, 94, 155, 156, 165
Martin
   Glenn Martin Company 39
   M-130 (Flying Boat Airliner) .... xii, xx, xxiv,
      xxvii, 35, 39, 46, 47, 50, 60, 61, 70, 71, 90, 111
Massachusetts Institute of Technology 12, 17
Massachusetts State College 17
Matto Grosso, Brazil 18
McCarty, William 18, 36, 63, 64, 80, 98, 99, 100
McKinley, Earl B. 5, 6, 7
   The Geography of Disease 6
McVay, Charles B. 178
Meier, Fred C. .... xxv, 6, 7
Mercator 52, 167, 168, 169, 171, 172
Mexico 4
Miami, Florida .... xvii, xxv, 17, 19
Micronesian Contractors 3, 70
Midway Island .... xiv, xxv, 3, 18, 28, 37, 39

Mili Atoll 23, 31, 94, 155, 165
Miller, Alton G. (*Glenn*) xii
Miller, William B. xiv
Miller, William T. 136
Milton, Massachusetts 17
Mississippi River 10
Mitchell, William 30
   Pacific Report (1924) 30
Mitsubishi .... xix, 29
   A6M2 Zero (Fighter) xix
   A6M2/5 Zero (Fighter) .... xix, xxii, 29, 31, 35, 65, 82, 83, 89, 102, 107
   A6M5 Zero (Fighter) 29
Mori, Taro 85, 109
Morrisey, Muriel (Earhart) 65
Murata, Shigeharu 28
Musick, Ed .... xxvi, xxvii, 25, 46, 89, 105, 107

N

Nagoya, Japan 69, 102, 170
Nanking, China 28, 29, 94
National Geographic Society 61
Nauroon, Robert 70, 85, 86, 93, 105, 109, 110, 151
Navigation
   Air Temperature 47
   Barometric Pressure 47, 48, 168
   Bearings 39, 43, 49, 52, 71, 98, 164, 174
   Cone of Silence 44
   Course Made Good 41, 57, 68
   Course Vector 47
   Current Set 53, 78, 81, 92
   Dead Reckoning .... xviii, xxvi, 36, 47, 53, 54, 57, 58, 59, 67, 68, 79, 94, 165, 167, 168
   Density Altitude 47
   Distance Made Good 41, 57, 100, 174
   Fix .... v, 36, 59, 62, 63, 64, 67, 71, 99, 100
   Great Circle 39, 52, 165, 167, 169, 171, 173, 174
   Greenwich mean Time (GMT) 48, 172
   Ground Speed (GS) 53, 58, 75, 171, 172
   Heading Vector 47
   Line of Position (LOP) 59, 62, 63, 64, 67, 75, 100, 155, 163, 164, 165, 172
   Radio Bearings 39, 43, 47, 52, 61, 71
   Running Fix 36, 39, 44, 63, 64, 67
   Sea Drift 47, 78, 92
   Single Line Approach (SLA) 39
   Sun Line 49, 50, 59
   True Air Speed (TAS) 47, 48, 50, 53, 54, 57, 58, 67, 71, 73, 75, 76, 90, 91, 97, 171, 172, 174
   Wind Direction (WD) 47, 168, 171, 172
   Wind Speed (WS) 168, 171, 172
   Wind Triangle (WT) 47, 48, 71
   Wind Vector (Speed and Direction) 47, 75
New Guinea .... xxiv, xxv, xxvi, 23, 37, 165
New Zealand .... xxvi, xxvii, 39, 46, 107
Newhall Pass, California 56

Newspapers
   The Bronxville Review-Press xiv
   The Cincinnati Post .... xii, xiv
   The Cincinnati Times-Star xiv
   The Harbin Nichi Nichi 105
   The Los Angeles Times .... xx, 70, 110
   The New York Times .... xiv, xviii, xx, 10, 23, 44, 56, 70, 75, 81, 82, 90, 92, 93, 105, 110, 177
   The Oregon Journal .... xiv, 3, 4, 84
   The Oregonian .... xiv, 3, 4
Nimitz, Chester 83, 107
Noonan, Frederick J. .... xii, xiv, xxv, xxvi, xxvii, 6, 12, 22, 30, 31, 39, 40, 41, 42, 43, 45, 57, 65
   Air Navigator .... xxv, 39
   Alcoholism 30, 40
   PAA Chief Navigator 30, 40, 45
   Round-the-World Flight 40
North, Oliver 31
Novak, Jack .... xii, 15

O

O'Sullivan, Anne (Jewett) xi
O'Sullivan, Jan .... xi, 17
Oakland, California .... xxi, xxiv, xxvi, 5, 24
Okumiya, Masatake 28
Orote Point, Guam 96

P

PAA Aircraft
   China Clipper (M-130) .... xxiv, 3, 25, 83, 105, 106, 110, 111
   Hawaii Clipper (M-130) .... v, xi, xii, xiv, xvii, xviii, xix, xx, xxi, xxii, xxiii, xxv, xxvii, xxviii, 3, 4, 5, 6, 7, 12, 14, 17, 18, 19, 21, 22, 23, 24, 25, 27, 29, 31, 35, 36, 40, 41, 42, 43, 44, 45, 46, 47, 48, 49, 50, 51, 52, 53, 54, 55, 56, 57, 58, 59, 60, 61, 62, 63, 64, 66, 67, 68, 69, 70, 71, 72, 73, 74, 75, 76, 77, 78, 79, 80, 81, 82, 83, 84, 86, 89, 90, 91, 92, 93, 94, 95, 96, 97, 98, 99, 100, 101, 102, 105, 106, 107, 110, 111, 112, 117, 118, 137, 150, 151, 152, 167, 168, 177, 178
   Hong Kong Clipper (S-42B) .... xxvii, 46
   Pan American Clipper (S-42) ... xiii, xxv, xxvii, 35
   Philippine Clipper (M-130) 41, 42, 56, 76
   Samoan Clipper (S-42B) .... xxvi, xxvii, 18, 25, 27, 29, 31, 46, 89, 94, 106
   Yankee Clipper (B-314) 22, 24
Pacific War .... v, ix, xix, xxi, xxii, xxiii, xxvi, 14, 27, 28, 30, 31, 35, 36, 37, 56, 65, 69, 70, 83, 93, 107, 187
Pagan, Marianas 82, 102, 170
Pago Pago, American Samoa .... xxvii, 18, 25, 31, 46
Palau Islands 38, 59, 60, 61, 62, 63, 64, 67, 69, 73, 75, 81, 89, 96, 97, 98, 101, 170
Palo Alto, California 19

Pan Am Aircraft
   Maid of the Seas..v, xix
Pan Am Historical Foundation xvii
Pan American Airways (PAA) .... v, xi, xii, xiv, xvii, xix, xx, xxi, xxiv, xxv, xxvi, xxvii, xxviii, 3, 5, 7, 9, 10, 12, 17, 18, 19, 21, 22, 24, 25, 30, 31, 35, 36, 37, 38, 39, 40, 41, 42, 43, 44, 45, 46, 47, 48, 49, 51, 53, 54, 55, 56, 59, 60, 61, 63, 64, 71, 72, 74, 75, 77, 78, 81, 83, 84, 85, 86, 89, 90, 92, 93, 94, 95, 97, 99, 101, 105, 106, 107, 109, 110, 111, 112, 113, 114, 115, 116, 117, 153, 154, 170
   Alaska Operations...12
   Caribbean Operations.......................12, 17, 18, 39
   Engineering Report............xvii, xx, 47, 49, 51, 53, 59, 77, 83, 86, 90, 110, 112, 113, 114, 115, 116
   Pacific Division.............................5, 18, 19, 24, 46
   PAMSCO..39
   Radio Makati (RDF) .... xviii, 35, 36, 43, 45, 49, 52, 57, 59, 61, 67, 68, 71, 90, 97, 98, 99, 101, 170, 171
   Radio Panay 35, 40, 43, 44, 48, 49, 50, 51, 52, 55, 59, 63, 64, 67, 68, 71, 74, 76, 89, 92, 94, 97, 98, 99, 100, 110, 170
   Radio Sumay............................31, 35, 41, 43, 49, 54, 59, 71, 95, 98, 99, 107, 170
   Trip 229 .... xii, xx, xxv, xxviii, 17, 18, 19, 24, 35, 47, 48, 84, 91, 95, 98, 101, 171
Pan American Foundation .... xvii, xix, 107, 110
Pan American World Airways (Pan Am) .... v, xiii, xiv, xvii, xix, 13, 70, 83, 105, 106, 107, 110, 153, 154
   Flight 103...v, xix, xxi, 107
Pangborn, Clyde xxiv
Parece Vela 81, 82, 93, 102, 170
Parker, Ivan..18, 95
Parker, Ruth..18
Pearl Harbor.. xiii, xxiii, 3, 18, 27, 28, 30, 37, 39, 71, 92, 187
Pearson Field 4
Pensacola, Florida..24
Philippine Islands (P.I.) .... xviii, xix, xxvii, 5, 6, 14, 23, 35, 37, 38, 41, 42, 43, 45, 46, 49, 55, 56, 61, 71, 73, 76, 77, 78, 101, 177, 187
Philipps, Michael xii
Phoenix Islands.. xxv, 31, 165
Pine Manor College 9, 12
Portland, Oregon..3, 4, 84
Portsmouth, Treaty of 37
Prange, Gordon W. .... xiii, 27, 28, 29
Pratt and Whitney .... xxv, 107
Prewett, Thomas xii
Priester, Andre A. .... xiv, xvii, 49, 50, 51, 53, 59, 77, 83, 90, 91, 92, 110, 112
Prymak, William xii
Puerto Rico 6, 23
Purdue University ....xi, 6, 23
Putnam, George P. .... xiii, xiv, 6, 23, 30, 31

Pyle, Ernie xxiii

R

Radar ....xxi, 39, 56
Radio .... v, xii, xiv, xvii, xviii, xxiv, xxv, xxvi, 18, 31, 35, 39, 40, 41, 42, 43, 47, 48, 49, 50, 51, 52, 55, 56, 59, 61, 63, 64, 65, 68, 71, 72, 73, 74, 76, 81, 89, 95, 96, 97, 98, 99, 100, 101, 155, 156, 157, 158, 159, 160, 161, 162, 163, 187
Radio Bearing *Mirror*...71
   Four Parameters 71
   Mirror Elements..................75, 79, 80, 90, 98, 99
Radio Direction Finder .... xviii, xxv, 31, 35, 36, 39, 40, 41, 42, 43, 44, 45, 49, 52, 57, 59, 61, 68, 71, 74, 76, 90, 97, 98, 99, 158, 170, 171, 174
   Adcock-type..............................39, 43, 45
   Loop-type............................40, 43, 74, 99
   Makati M-101T..................49, 52, 67, 68, 99, 171
Radio Transcripts....................................157, 158, 162
Radio Wave Propagation
   Ground Waves 156, 157
   Line of Sight 156
   Sky Waves 156
Radio-Deception...v, 35, 71
Rain Static 50, 51, 52, 55, 89
Reineck, Rollin C. xii
Richter Library xvii
Rising Sun with Rays 35, 37
Robinson, Lynette (Kennedy)..........................xi, 5
Rockefeller Foundation 5
Ronin (masterless Samurai)..................................27
Roosevelt, Eleanor ....xxv, xxvi
Roosevelt, Franklin D. .... xxiii, 77
   Executive Order 7959 77
Roosevelt, Theodore..37
Roseberry, Cecil xiii
Rota Island, Marianas 42
Roxbury School 10
Russia 18, 37, 71

S

S.S. *Lurline*...18
Sabotage ....xix, xx, 22, 45, 61, 89, 105, 106, 138
Sailings .... xviii, 167, 168
   Celestial Navigation...168
   Constant (Pi) 171
   Contemporary Flight Data 171
   Dead-Reckoning 79, 168
   Derived Functions for BASIC...........................174
   Distance Conversion 171
   Dynamic Flight Equations 174
   Function Designators 171
   Geographic Positions 170
   Great Circle..169, 173
   Great Circle Equations 173
   Instrument Equations 174
   Interdependent Equations.................................175

    Latitude/Longitude Conversion 171
    Loxodrome..168
    Mathematical Operators..175
    Mercator Projection 169
    Pilotage 168
    Rhumbline..169, 173
    Rhumbline Equations...173
    Temperature Conversion.......................................171
    Time Conversion..172
    True bearing Angle 168
    Unit Abbreviations...171
    Wind Triangle Equations 172
Saipan ....ix, 31, 38, 42, 63, 64, 65, 66, 67, 68, 69, 83, 99, 146, 170, 177
Sakhalin Island 37
Salzman, Phil C. ....xiv, 136
Samara, Russia..18
Samurai... xiii, 27
San Bernardino Strait......................40, 43, 45, 49, 55
San Francisco..........................xi, xiii, xxiv, xxvii, 4, 9, 18, 41, 42, 56, 57, 91, 95, 154, 158, 161
Sangley Point, P.I. 45
Santa Barbara, California..17
Santa Cruz, Battle of...28
Sarah Lawrence College 19
Sasebo, Japan..29
Sauceda, Jose M. ....xxii, 18, 24, 93, 98
Scammell, Henry xiv
Schiff, Judith... xii, 15
School of Tropical Medicine (Puerto Rico).............6
Seattle, Washington 17
Seckman, Michael xii
Sensei (master) ....xi, 27
Shanghai, China..25, 28, 29
Sherman, William Tecumseh..27
Sikorsky.......xxvii, 19, 22, 38, 39, 46, 73, 89, 95, 106
    S-42 (Flying Boat Airliner).........................xxvii, 38, 39, 73, 89, 95, 106, 107
    S-42B (Flying Boat Airliner) ....xxvii, 19, 46
Silverman, Bob xii
Smith, Richard K. xiv
Smith, Sumpter xiv
Soma, Arthur xii
South America 18, 22
South Lee, Massachusetts...17
Southern Route 41, 42, 43, 44, 45, 49, 51, 55
Soviet Communism ....xxiv, xxvi, 31
Soviet Siberia................................. xxiii, 27, 28, 38
Soviet Union .... xxiii, xxvi, 27, 28, 31, 35
Spanish-American War..37
Stanford University...24
Steigmann, Jerome.. xii
Strategic Plotting Chart 5050-3 ....164, 165
Streeter, Hope (Wyman)............... xi, xiv, xix, xx, xxi, 9, 10, 11, 12, 13, 14, 15, 16, 105
Sutherland, Richard Kerens .... xi, xii, xiv, xix, xxi, xxii, 9, 12, 13, 14, 15, 69, 83, 89, 105

Sutter, William L. ....xii, xxv
Swan Island Airport (Portland)................................4

T

Tail-hookers...30
Takao, Formosa 60, 73, 74, 76, 81, 91, 92, 94, 97, 100, 101, 170
Tanapag Harbor, Saipan 63, 99
Tanapag, Saipan..............................63, 65, 99, 170
Tarn, E. A. 106
Tatum, Thomas B. 17, 18, 95
Terletsky, Leo 17, 18, 19, 24, 55, 78, 96, 97, 100, 107
Terletsky, Saretta (Bowman)................................18
Thach, Robert 106
Thompson, Gerald 161
Tientsin, China..106
Tinian Island, Marianas 42
Togo, Heihachiro...35, 37
Tojo, Hideki...ix
Tokyo Rose...65
Tokyo, Japan....................................... xviii, 7, 15, 38, 60, 61, 62, 65, 81, 82, 83, 93, 101, 102, 177
Towers, John Henry..............xix, xxii, 14, 83, 89, 107
Traynor, Josh xi
Trippe, Juan T. .... xiii, xix, xxi, 9, 12, 14, 83, 84, 105, 106, 107
Truk Atoll .... v, xi, xx, xxi, xxii, 23, 35, 36, 38, 67, 69, 70, 72, 73, 75, 76, 82, 85, 86, 89, 93, 95, 98, 99, 100, 101, 102, 109, 110, 140, 142, 143, 145, 146, 147, 149, 152, 170
    Amelia Earhart Law.............................. xxii, 70, 86
    Gibraltar of the Pacific.................................35, 38
Tsushima Straits, Battle of.................................37
Tucson, Arizona...9
Tutuila, American Samoa xxvii

U

U.S. Air Force...14, 92
    Air Force Museum (WPAFB)................... xxii, 83
U.S. Army
    Medical Corps...6
    Postal Unit (Saipan, 1944)................................65
    Signal Corps... xii, 15
    USAT Meigs...53, 57, 74, 77, 78, 81, 89, 91, 92, 94, 97, 100, 101, 170
U.S. Army Air Corps....................................... xxvi, 4, 5, 23, 28, 53, 89, 92
    321st Observation Squadron............................4, 84
U.S. Bureau of Plant Industry..6
U.S. Coast Guard ....xxiv, 7, 31, 155, 156, 158, 161, 187
    Loran Transmitting Station 7, 156, 157
    USCGC Itasca .... xii, xiv, xxiv, xxv, xxvi, 31, 155, 156, 157, 158, 159, 160, 161, 162, 163, 164, 165
    USCGC Taney 158

U.S. Constitution
  First Amendment iv
  Ninth Amendment iv
U.S. Department of Agriculture .... xxv, 6
U.S. Department of Commerce .... xiv, xviii, xix, xxiv, xxvii, xxviii, 21, 22, 60, 77, 81, 89, 94, 110, 120, 136
  Board .... xviii, xix, 73, 77, 78, 91, 92
  Inquiry Board .... xviii, 77, 91
  Secretary of .... xviii, 21, 77, 106
U.S. Department of Interior
  Trust Territory 86
U.S. Department of State 81, 82, 93
U.S. Department of Treasury .... xx, 66, 83, 84
U.S. Department of War
  Secretary of War 65
U.S. Federal Bureau of Investigation .... xxvii, 22, 105, 106, 107
U.S. Gold Certificates 35, 83, 92
U.S. Marine Corps .... xii, xix, 12, 39, 41, 45, 61, 65, 66, 83, 89, 95, 107
U.S. National Air and Space Museum 29
U.S. National Archives .... xxv, 105, 158
U.S. National Research Council 6
U.S. National Transportation Safety Board ... xii, xvii, 49, 77
U.S. Naval Academy .... xi, 24
U.S. Naval Academy Alumni Association
  Shipmate Magazine .... xi, xii, xiv, 21
U.S. Naval Institute Press 28
U.S. Navy .... xxii, xxv, xxvii, 17, 24, 25, 27, 28, 30, 35, 36, 37, 39, 41, 45, 53, 61, 65, 69, 81, 82, 83, 84, 93, 101, 105, 106, 107, 110, 164, 165, 178
  Asiatic Fleet 53
  Bureau of Aeronautics 106
  Hydrographic Office 164
  Submarine Mafia 69
  USS *Augusta* 25
  USS *Avocet* xxvii
  USS *Indianapolis* 81, 178
  USS *Langley* 83, 107
  USS *Missouri* .... xii, 15, 83
  USS *Panay* .... xxi, xxvi, 25, 27, 28, 29, 93, 94
  USS *Saratoga* 22, 83, 107
U.S. Navy Reserves 12, 17, 19, 24, 25, 35
Ulithi Atoll .... v, xviii, 35, 36, 67, 71, 72, 73, 76, 90, 95, 96, 97, 98, 99, 100, 170, 178
Ulm, Charles xxiv
Ultimate M.I.A.'s .... v, xii, 35
Umezu, Yoshijiro .... xii, 15
Underhill, John xii
Unimakur Mountain, Dublon 85, 101
Union Air Terminal, Los Angeles 56
United Airlines 4, 5, 56
United Press 78, 81
University of Brussels 6

University of Michigan 6
University of the Philippines 6
Unsworth, Michael E. xiv

V

Vancouver, Washington 4
Vidal, Eugene xxiv
Virden, Stan xii

W

WACO 18
Wake Island .... xiii, xxv, 18, 37, 39, 41, 73, 76, 95, 170
  Night Landing, 1935 76
Walker
  Recovery Authorization 26
Walker, Mark Anderson .... xi, xii, xxi, xxv, xxviii, 17, 18, 19, 21, 22, 23, 24, 25, 26, 35, 89, 93, 100
War Crimes .... v, xxii, 36, 86, 93, 110
Washington, D.C. .... xxi, 29
Waterbury, Connecticut 10
Wecker, David xii
Weideman, Christine .... xii, 15
White House .... ix, 66, 177
Wichita falls, Texas ..... xiv, xxi, xxviii, 19, 21, 22, 24
Wiley, Barbara xi
Williams, Clarence 153, 165
Wilson, Marjory (Kennedy) .... xi, 5
Wilson, Woodrow 37
World War I .... xiv, xxiii, 4, 6, 7, 12, 37, 61, 71, 83, 84, 107
World War II .... xii, xiv, xxiii, 7, 12, 65, 69, 83, 107, 109, 110, 147, 148
Wyman
  Appeal for an Inquiry, 1980 13
  Recovery Authorization 16
  Sales Trip Itinerary, 1938 11
Wyman, Corydon 9, 12
Wyman, Edna Benton (Wake) .... xix, 9, 10, 11, 12, 14, 15, 69, 105
Wyman, Edward Earle .... xi, xix, xxi, 7, 9, 10, 11, 12, 14, 18, 35, 69, 70, 84, 92

Y

Yale University .... xii, xiv, 10, 12, 14, 15
  Library .... xii, xiv, 15
  Yale Classbook, 1916 .... xiv, 15
Yamamoto, Isoroku 30, 31
Yangtze River .... xxvi, 25, 28, 93
Yokohama, Japan 61, 107
Yokosuka, Japan .... xix, 83, 89, 102
Yucatan, Mexico 19

Z

Zellhart, Paul xii
Zembsch, Lawrence 61, 62
Zero Hour (Propaganda Radio Program) 65

7. ABOUT THE AUTHOR

Charles N. Hill was only seven months old when Japanese naval aviators attacked Pearl Harbor, but soon after, with one of his two uncles serving in North Africa and, later, in Italy, and the other held as a prisoner of war by the Japanese in the southern Philippines, the two concurrent wars came to serve as a very personal backdrop to his early years.

By the summer of 1945, it was customary for neighborhood boys to play "war," almost incessantly. They seldom fought the Nazis in their games, and, because he was younger and smaller than many of the boys, Charles spent most of his time, in play, as a "Jap." Late in 1945, Charles' mother received word that her brother, Major Robert Nelson, had starved to death aboard a Japanese POW *maru*—one of the infamous hell-ships—en route to Japan, on January 20, 1945, two and a half years after Roosevelt had ordered the surrender of the Philippines. Charles recalls that his mother's reaction to the notification of her brother's death actually terrified him. He never played a "Jap," after that, nor was the war ever discussed at his mother's table, even as it is forbidden in Japan, even today.

But reading was not forbidden, so Charles immersed himself in a study of the war, at one time owning a library of four thousand volumes, half of it pertaining to the Pacific War. The catharsis washed away his learned hatred and replaced it with a more objective interest. In 1974, he became intrigued by the Klass/Gervais book, *Amelia Earhart Lives*, and began his own pursuit of the lady whom he then regarded as a victim of the Japanese. Today, he knows that the Japanese were *her* victims, but his pursuit of Earhart continues, and the 1938 Clipper hi-jacking, he believes, will eventually open a "back door" into the "imagination-staggering" saga of the woman whom he is convinced was "twice a traitor."

But his quest has not been without hazard or adversity. In November, 1989, Charles was invited to speak at Purdue University, at a symposium of the Amelia Earhart Research Consortium, a shortlived organization of the best informed private Earhart researchers in the U.S. Upon his arrival at Purdue, Charles learned, to his surprise, that Joe Gervais, of all researchers, was not scheduled to attend. Contacted by phone, Joe firmly told Charles, "Pack your bags and get out of Purdue," and explained that the "silent partner" of the AERC was a C.I.A. operative and that the symposium was serving as a C.I.A. "sting" operation, aimed at determining who knew what about Amelia Earhart. Charles thought that Joe's charge was too bizarre to be true, and so he foolishly ignored Joe's warning.

Two weeks later, Charles was advised by his employers that he had been declared to be a "security risk" and that he would have to be terminated as soon as he had completed a then-current project. Since then, both he and his wife have found it nearly impossible to secure regular employment, yet, the idea that they might be victims of a *U.S. Government blacklist* seemed as absurd as Gervais' claim of a C.I.A sting against Earhart researchers. But, in 1996, the former "silent partner" of the AERC advised Charles that he had been employed by the C.I.A. in 1989 and that the symposium had, indeed, served as a covert C.I.A. operation: *the blacklist was quite real and continues to be—an oppressive reality.*

Charles was graduated, in 1972, *cum Laude*, with Honors in English, from the University of Cincinnati, but draws upon Coast Guard electronics training, whenever he can, for his livelihood. He lives with his wife, Carol, and their sons, Ian and Charles, in the shadows of the American Dream, hoping, eventually, to complete his next book, *Twice a Traitor*.

Printed in the United States
2457